I0063851

MACHINERY'S REFERENCE SERIES

EACH NUMBER IS ONE UNIT IN A COMPLETE LIBRARY OF MACHINE DESIGN AND SHOP PRACTICE REVISED AND REPUBLISHED FROM MACHINERY

DESIGNING AND CUTTING CAMS

SECOND EDITION—REVISED AND ENLARGED

CONTENTS

49-55 Lafayette Street, New York City

CHAPTER I

THE DRAFTING OF CAMS*

A cam is a device for converting circular into reciprocating motion. It generally consists of a disk having an irregular face that acts as driver of a follower in contact with it, or else of a groove cut in a flat or curved surface. Cams are very useful adjuncts to many forms of machines, as by their aid various complex and complicated movements may be obtained that were otherwise impossible. Their use is, however, attended with some objections of a character serious enough to warrant the substitution of some other method of arriving at a desired result when such other method is available. Among these objections may be mentioned the considerable amount of friction, producing wear, and the noisy action of cam movements. Despite these objections, cams have a wide use and are employed in many familiar machines.

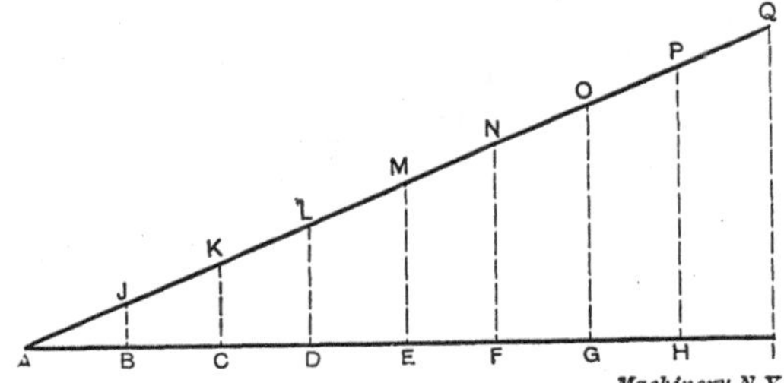

Fig. 1. Diagram graphically showing Motion Imparted to Follower by Cams in Figs. 2 and 3

Harvesters, printing presses, sewing machines, looms, and steam-valve mechanisms are a few of such machines to which cams contribute part of the action. The more complicated forms of automatic machinery, automatic screw machines, for instance, depend largely upon the aid of cams. The various machines used in the manufacture of shoes are also good examples of this class.

Laying Out a Cam for Uniform Reciprocating Motion

The knowledge of laying out cams is simply and easily acquired. The laying-out of a heart-shaped cam will serve as an illustration of the general method. This cam is used to convert circular motion into uniform reciprocating motion. Let it be required to lay out a cam that will move a follower with uniform velocity through a throw of 1½ inch. This action may be graphically shown by the aid of a diagram, Fig. 1. The action of but one-half the complete movement need be considered, as the return of the follower is along a curve similar to that occasioning the rise. Therefore, let *A I*, a line of indefinite length,

* MACHINERY, March, April and December, 1897.

represent one-half a revolution of the cam. At *I* draw the perpendicular *I Q* equal to the extreme throw, in this case 1½ inch. As the rise of the follower is to be uniform, this action may be shown by a straight line connecting *A* and *Q*. Divide the line *A I* into any number of equal parts, say eight, and erect perpendiculars at the points of division. The point *E* will then represent one-quarter revolution of the cam, and the distance *E M* will represent the throw at that point. In the same way the distance *C K* represents the amount of throw at one-eighth revolution, the distance *G O*, the throw at three-eighths revolution, and so on for the other perpendiculars.

To lay out the cam curve, describe about *X*, Fig. 2, as center any semi-circle, *a e i*. Divide this semi-circle into the same number of

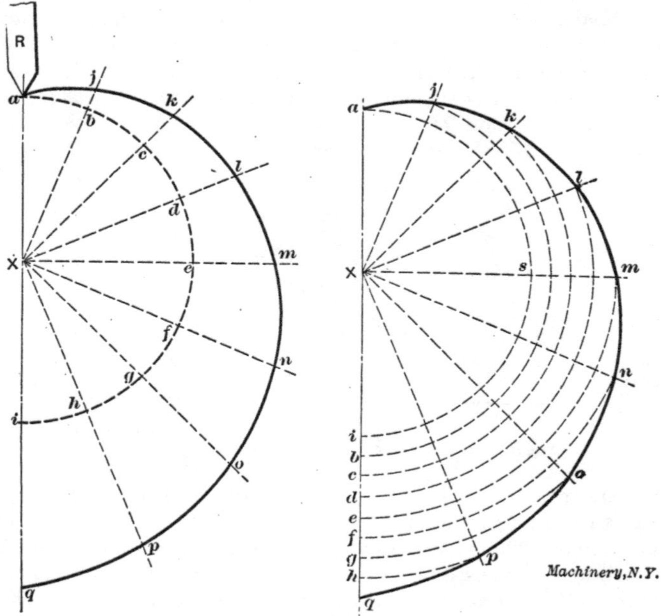

Figs. 2 and 3. Lay-out of Uniform Motion Cams

equal parts into which the line *A I* was divided. Connect these points of division with the center *X*, and extend the lines indefinitely beyond the semi-circle. On *X b*, make *b j* equal to *B J*, on *X c*, make *c k* equal to *C K*, and so on, extending each radius a distance equal to the corresponding perpendicular in Fig. 1. Then through the points *a*, *j*, *k*, *l*, etc., draw a smooth curve. This curve is one-half the required cam curve. By drawing a similar curve to the left of *a* the cam curve is completed. By rotating the cam about the center, *X*, the follower, *R*, would be forced to rise, with uniform velocity, through a distance of 1½ inch. During the second half of the revolution it would fall uniformly, by aid of gravity or a spring, to the initial point *a*.

Alternative Method of Laying Out Cam Curve

Another way of laying out the same cam curve is as follows: Draw any semi-circle, *a s i*, Fig. 3, and extend the diameter on one side a distance *i q* equal to the required throw. Divide *i q* into any number of equal parts, as at *b*, *c*, *d*, etc., and divide the semi-circle by the same number of radii equally distributed. With *X* as center and a radius

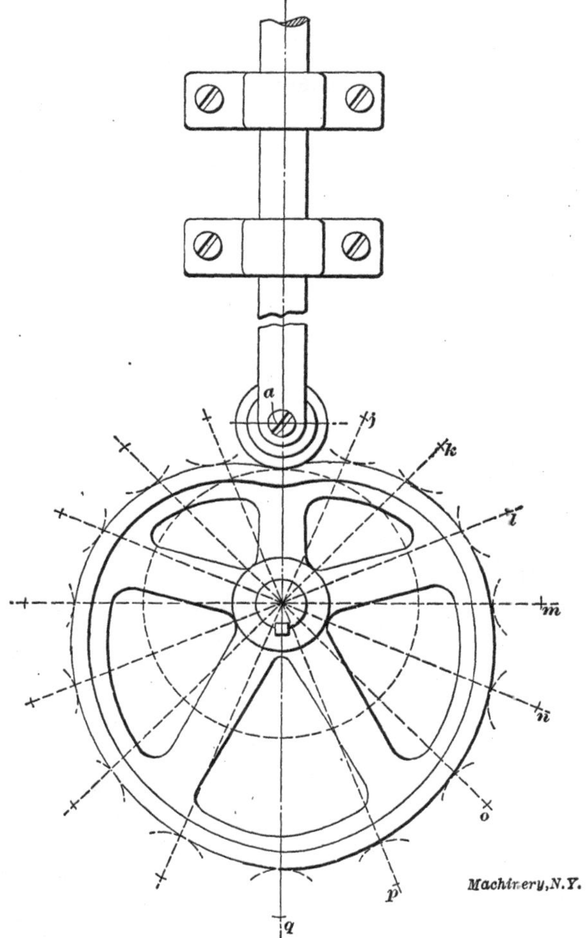

Fig. 4. Cam with Roller Follower

equal to *X b* describe an arc cutting *X j* at *j*. With the same center and radius equal to *X c* describe an arc cutting *X k* at *k*. Continue this process through the points *d*, *e*, *f*, etc., thus obtaining the points *l*, *m*, *n*, etc. The latter are points on the required curve.

The excessive friction of a pointed follower such as that shown at *R* necessitates the employment of a follower that will reduce the amount of friction to a minimum. A small roller meets this requirement. If a roller is employed as a follower the problem of laying out the cam curve becomes modified. A roller traveling along the curves shown in Figs. 2 and 3 would not impart to the follower-rod the desired uniform rise and fall. The variation would be but slight, yet sufficient to merit consideration where accuracy is desired.

Cams with Roller Followers

Fig. 4 represents a heart-shaped cam of the same dimensions as in Figs. 2 and 3, but with a roller follower. It is the path of the center

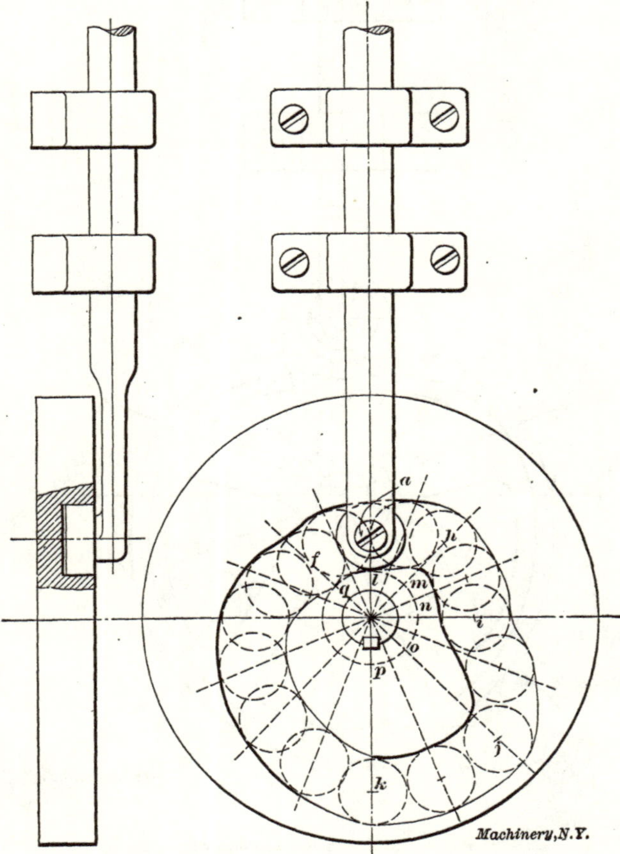

Fig. 5. Positive Action Cam for Variable Motion of Follower

of this roller that requires the first consideration, as the position of this center regulates the throw. Therefore, the position of the center of the roller at various intervals in the rotation of the cam must be determined. This may be done by adding to each of the distances *J B*, *K C*, *L D*, etc., in Fig. 1, the radius of the roller, and thus obtaining the points *j*, *k*, *l*, etc. With these points as centers and with radii equal to that of the roller, describe arcs. A curve drawn tangent to these arcs is the required cam curve.

This cam depends upon the action of gravity, or a spring, to keep the follower in contact with the driver. It can be made positive in action by the use of two followers placed at the extremities of the diameter of the cam, or by drawing curves tangent to both the top and bottom of the follower roller in its various positions, and the two curves taken as the boundaries of a groove cut into the metal. A familiar application of the use of a heart-shaped cam may be found in the bobbin-winder of the domestic sewing machine. The thread is fed to and fro at a uniform rate, the follower of the cam acting as a guide for the thread. The action is made positive by the employment of two follower rollers.

Positive Action Cam for Variable Motion of Follower

The latter method of laying out a positive motion cam referred to above is more clearly shown in Fig. 5. A variable motion is here substituted for the regular motion of the heart-shaped cam. Let it be required to lay out a positive motion cam that shall impart to the

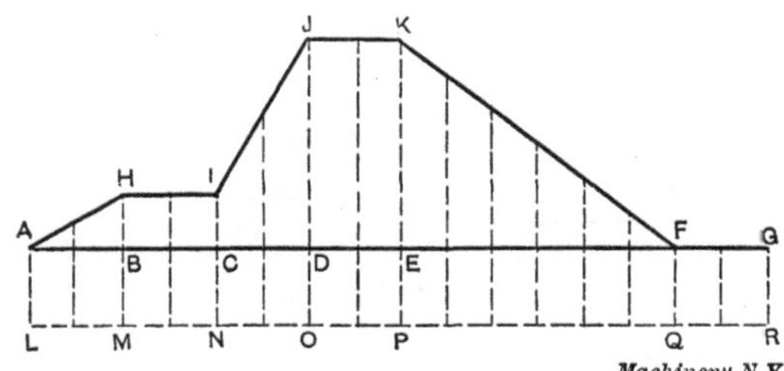

Fig. 6. Diagram of Motion Imparted to Follower by Cam in Fig. 5

follower the following action: A uniform rise of ¼ inch during the first eighth of a revolution; no action during the next eighth; a uniform rise of ¾ inch during the third eighth; no action during the fourth eighth; a uniform fall of 1 inch during the next three-eighths of the revolution; and no action during the last eighth. The action is graphically shown in Fig. 6. Let *A G* represent one complete revolution of the cam; *B*, the first eighth; *C*, the second; *D*, the third; *E*, the fourth; and *F*, the seventh. The problem calls for a uniform rise of ¼ inch during the first eighth. Therefore, from *B* draw the perpendicular *B H*, ¼ inch in length, and join *A* and *H*. As there is to be no action during the second eighth, draw *H I* parallel to *B C*; that is, the follower will be the same distance from *A G* at *I* that it was at *H*, and therefore the follower will not have been acted upon. During the next eighth revolution the follower is required to move ¾ inch. As it has already moved ¼ inch, the sum of these two distances is the length *D J*. As this rise is to be uniform, a straight line is drawn joining *I* and *J*. No action during the fourth eighth is shown by drawing *J K* parallel to *D E*. A uniform fall of 1 inch during the next three-eighths of the revolution is shown by joining *K* and *F*, and the period of rest during the last eighth revolution is shown at *F G*. The line *A L* is equal to the radius of the roller, and by drawing the line *L R* parallel to *A G*, the distance of the center of the roller from the base circle may be taken directly for any radius of the cam.

To lay out the cam from the diagram, draw any base circle *l n p*, Fig. 5, and divide it into the same number of equal parts into which the line *A G* is divided, *viz.*, sixteen. Through these points of division

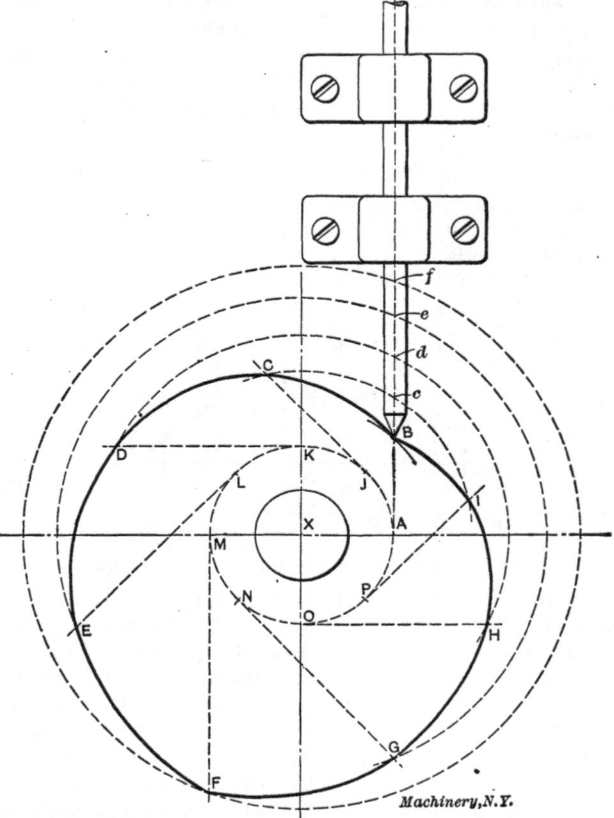

Fig. 7. Cam with Follower having Line of Action Eccentric with Cam Axis

draw radii and extend them indefinitely. Upon these radii take $la = LA$, $mh = MH$, $ni = NI$, etc., thus determining the positions of the center of the roller at the various intervals. Sketch in the outline of the roller in its different positions, and draw curves tangent to these outlines.

Line of Action of Follower Eccentric with Cam Axis

In the cams previously considered, the line of action of the follower passes through the center of the cam-shaft. When the line of action of the follower passes to either side of the center of the cam-shaft, as in Fig. 7, a different method of laying out the cam curve becomes necessary. Assume that the requirements and conditions are the same as in Fig. 2, excepting that the line of action of the follower shall be one inch to the right of the center of the cam-shaft. Draw the indefinite line *X A* passing through the center of the cam-shaft. One inch to the right of *X* draw the line of action *A f*, of the follower, perpendicular to *X A*. Let *B* be the lowest position that the follower is to assume, and let *f* be the highest. Divide the throw, *B f*, into any number of equal parts, as at *c*, *d* and *e*. Through *A* describe a circle with *X* as center. Divide this circle into twice the number of equal parts into which *B f* is divided. From each of these points *J*, *K*, *L*, etc., draw tangents to the circle. Then, with *X* as center, describe arcs through *c*, *d*, *e*, and *f*. Where the arc *c* cuts the tangents from points

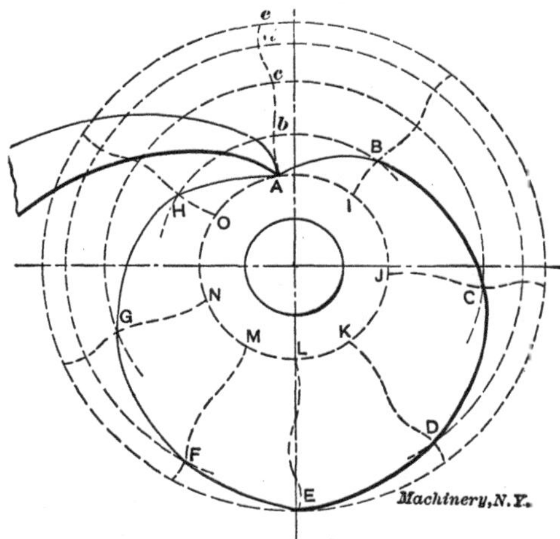

Fig. 8. Cam and Follower both having Variable Motion

J and *P*, as at *C* and *I*, are points on the desired curve. Where the arc through *d* cuts the tangents from *K* and *O*, as at *D* and *H*, are also points on the curve. The points *E*, *F*, and *G* are obtained in a like manner.

Cams with Pivoted Followers

The problem in Fig. 2 may be further modified by having the follower pivoted instead of acting in a straight line. In this case, the line of action becomes the arc of a circle. Problems of this nature may be solved by substituting for the straight line of action shown at *i q*, Fig. 3, an arc which shall represent the path of the follower. This arc of action takes the place of all the various radii in Fig. 3, and the points *b*, *c*, *d*, etc., serve as a series of initial points from which to swing concentric arcs to intersect the various positions of the arc of action of the follower. The method is analogous to that in Fig. 3. In Fig. 29 this method is applied to a cam of un-uniform motion.

Cams and Followers both having Variable Motion

The rotation of the driver has thus far been considered as uniform, and the action of the follower either uniform or irregular. A case will now be considered wherein both the action of the driver and that of the follower is irregular. In Fig. 8, let the unequal divisions into which the base circle *A J L* is divided by the points *A*, *I*, *J*, etc., represent spaces traversed by the driver in equal periods of time. That is, if it takes the driver one second to rotate through the arc *A I*, it will take the same time to rotate through the larger arc *I J* or the smaller arc *L M*. Again, let *A e* represent the irregular path of the follower and the points *b*, *c*, *d*, and *e* its position at certain equal intervals of time, say one second. The number of divisions made in the path

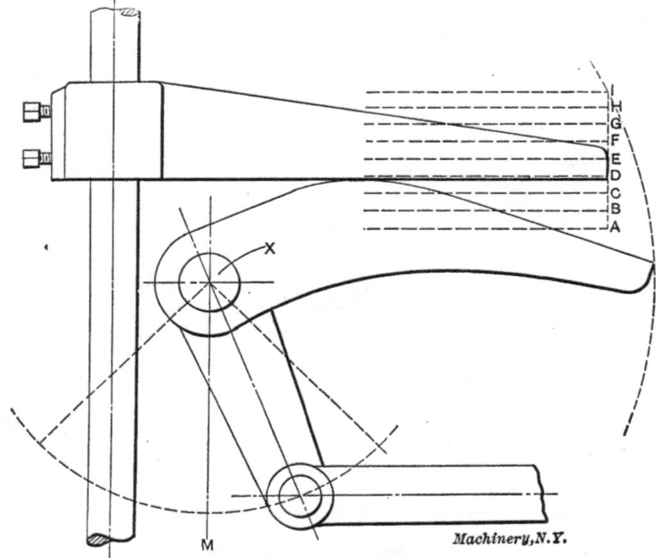

Fig. 9. Cam with "Flat-footed" Follower

of the follower should correspond with the number of divisions into which one revolution of the driver is divided. The points *B*, *C*, *D*, etc., of the cam curve may be found by the method of intersections explained in Fig. 3. This problem is of a general nature and is universally applicable to problems involving a disk driver and a follower other than a flat-footed one.

The "Flat-footed" Follower

A familiar example of a flat-footed follower is afforded by the toe-and-lift mechanism used to actuate the engine valves of side-wheel steamers. The "lift" or "wiper" is pivoted upon a rock-shaft which is caused to oscillate by an eccentric placed upon the paddle-wheel shaft. In Fig. 9, let the arc through which the rock-shaft swings equal 90 degrees—45 degrees on either side of the vertical—and let the "toe" rise and fall with uniform motion through 1½ inch. It is required to design the upper face of the lift to give the desired throw.

Divide the throw, *A I*, into any number of equal parts, say eight, and locate the center of the rock-shaft, as *X*. Upon a piece of tracing paper draw a quadrant, *x h k*, Fig. 10, *x k* being equal to one-half the throw of the eccentric, say 3 inches. Draw *x l* at 45 degrees to *x h*, and *k l* at right angles to *x k*. Through the point of intersection, *l*, and with *x* as center, describe the arc *l m*. The arc *k h* then represents a quarter revolution of the eccentric, and the arc *l m* the corresponding angular movement of the rock-shaft crank. Divide the arc *k h* into the same number of equal parts into which the throw of the toe was divided, *viz.*, eight. Through these points of division draw lines parallel to *x m*, intersecting the arc *m l* in the points *n*, *o*, *p*, etc. From these points draw radial lines. Now, while the eccentric is moving through a quarter revolution with a uniform motion, as shown

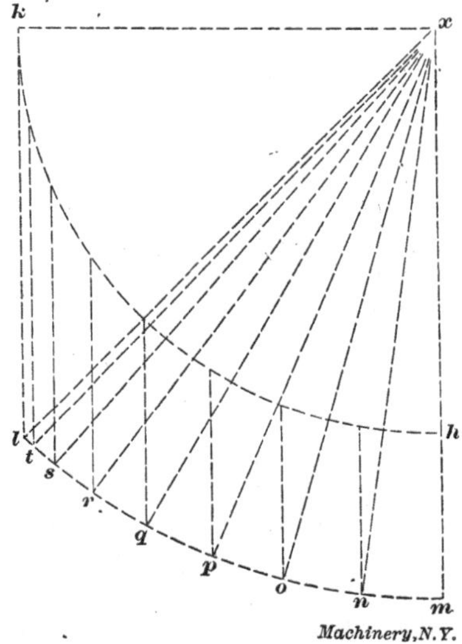

Fig. 10. Lay-out for Cam with "Flat-footed" Follower

by the equal division of the arc *h k*, the center line of the rock-shaft crank will assume the corresponding positions shown by the radial lines.

Place the tracing in Fig. 10 upon Fig. 9 so that *X* and *x* coincide, and the line *x m* falls upon the line *X M*. Then draw upon the tracing-paper the position of the line *A*. Rotate the tracing-paper about *X* until *x n* coincides with *X M* and draw the position of the line *B*. Again rotate about *X* until *x o* coincides with *X M* and draw the position of the line *C*. Continue this process until the positions of the lines *D*, *E*, *F*, etc., are located. A curve drawn tangent to the lines thus obtained is the required cam curve. This latter procedure is not shown in the cuts. The use of tracing-paper for laying out cam curves, as here exemplified, is applicable to the laying-out of a variety of such curves. The tracing may be made to assume different positions of either the driver or follower, and their relation shown at any desired interval during their action.

In work dealing with cam curves there are some factors of a practical nature that must be considered, one of which may be here stated, as applying directly to the problem of the toe-and-lift. This factor is the easement of cam action to prevent jerking. The action as drawn in Fig. 9 has too abrupt a beginning and ending, and should be modified by an easement curve at both these points of action. In any action that tends to jerkiness, a smoother motion may be obtained by slightly modifying the curve at the offending point.

Cams with Double Contact

In the drawings of cams thus far shown, there has been but one point of contact between the driver and follower. Positive motion is often obtained by having two points of contact. Cams having two such

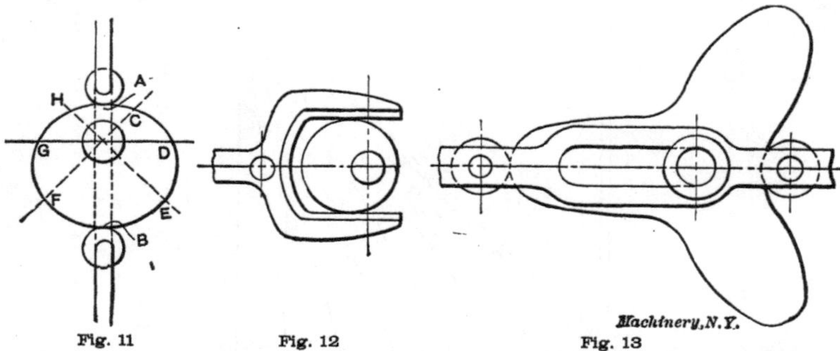

Fig. 11 Fig. 12 Fig. 13

points of contact are subject to certain limitations. For instance, in Fig. 11, if *A* and *B* are two points of contact of the follower, and are a constant distance apart, and the curve *A D B* be any assumed curve of one-half revolution of the cam, the curve of the remaining half revolution is limited to a curve complementary to *A D B*. That is, the distances *C F*, *D G*, and *E H* must equal the constant *A B*.

If it is desired to have an independent movement throughout the entire revolution of the cam it will be necessary to have two cams placed one upon the other, one point of contact of the follower bearing upon the second cam. In this case, having assumed any curve for one of the cams, the other cam must be made complementary to the first, the constant distance apart of the points of contact forming the basis for the calculation. Forms of double contact cams are shown in Figs 12 and 13. Fig. 12 is a rocker cam, and Fig. 13 is a tri-lobe cam giving tnree reciprocating motions to the follower for each revolution of the driver.

Cylindrical Cams

Fig. 14 illustrates a method for laying out cylindrical cams. Let *g d a* be the plan, and *H a'* the development of the cylinder shown in elevation at *KA*. Divide the plan into any number of equal parts as at *a*, *b*, *c*, etc., and project these points of division upon the front elevation of the cylinder as the elements *A*, *B*, *C*, etc. On the developed surface these elements appear as *a'*, *b'*, *c'*, etc. Upon the development, lay out the desired action, as in Figs. 1 and 6, avoiding or easing all sharp corners. Suppose *m l p* to be such an action. This curve will then represent the path of the center of the follower. Let *L* indicate the center of the follower. Then, as the cylinder is rotated about its axis, the point *L* moves to and fro a distance *L I*, and with an irregular motion dependent on the form of the curve *m l p*. The projection of this curve upon the elevation of the cylinder is shown at *L m*.

The form of the roller-follower may be either cylindrical or conical; the question of the shape of the follower has been treated more completely in Chapter V. In laying out the cam practically, the outline of the groove may be drawn by the method shown in Fig. 5, that is,

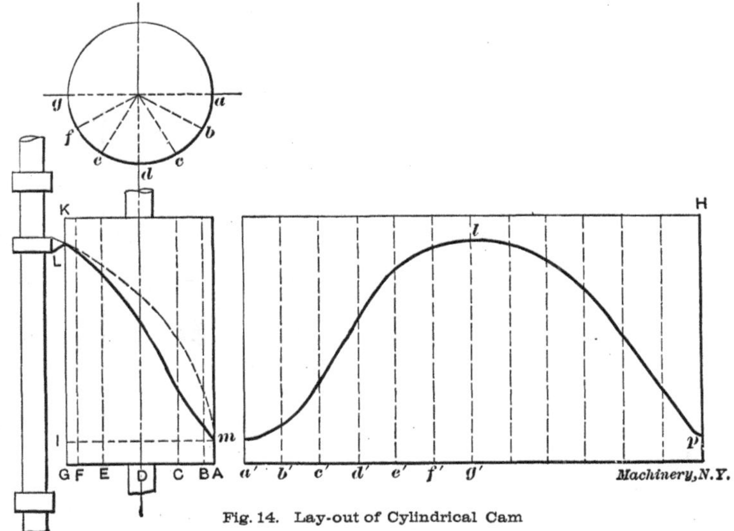

Fig. 14. Lay-out of Cylindrical Cam

by drawing curves tangent to the various positions of the roller, and then, by winding the drawing about the metal cylinder blank, any number of points of the groove may be located with a prick-punch; or, the drawing may be made directly upon the surface of the cylinder.

The method for laying out a conical cam is similar in principle to that for laying out a cylindrical cam, and is easily deduced from the latter.

Laying Out a Cam for Shifting Planer Belts

The following problem in machine design is one of a series given to the students in mechanical engineering at Cornell University. It furnishes a good example of the method of reasoning applied to practical problems in mechanics, and is also an interesting problem in quick-return motions. The problem calls for the designing of a device for automatically shifting the belts of a planer. The driving shaft has a fixed pulley of wide face carrying two belts. The driven shaft has two sets of a loose and a fixed pulley. One set, smaller than the other, is driven by a crossed belt, and its shaft therefore rotates in a direction opposite to that of the driving shaft. The larger fixed pulley drives the planer while the tool is cutting, and the smaller fixed

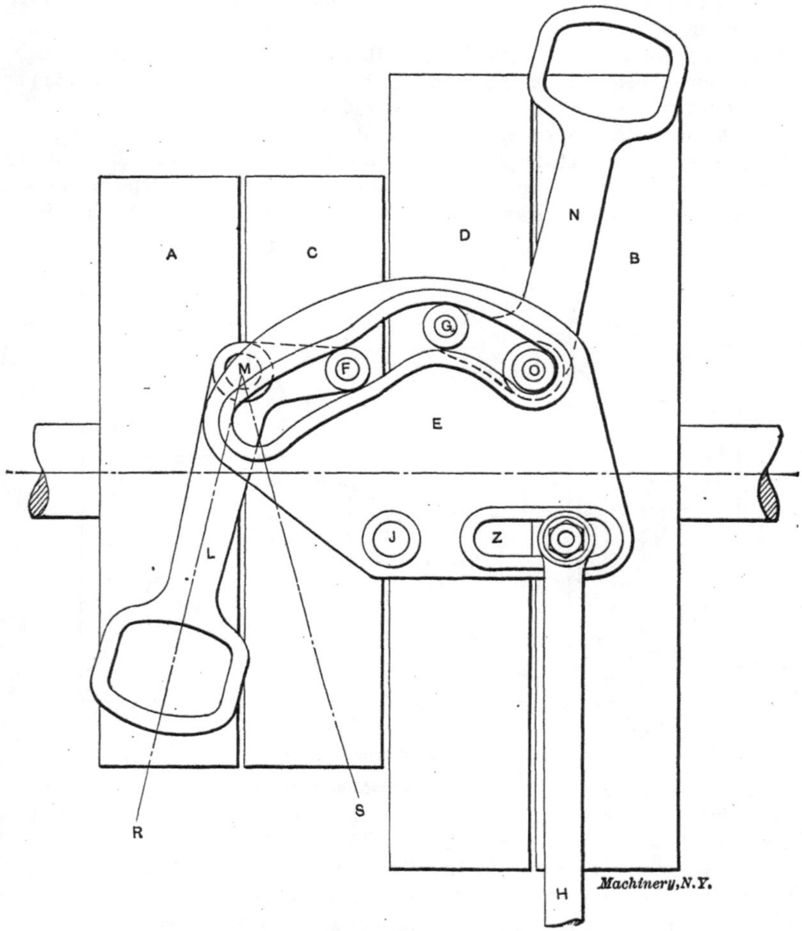

Fig. 15. Arrangement for Automatically Shifting Planer Belts

pulley causes a quick return of the tool while no work is being performed.

The shifter should be placed near the driven pulleys so as to operate each of the belts at its point of approach to its pulley, and to operate each belt separately. The shifter must also be operated automatically by the to-and-fro motion of the bed of the planer, and be capable of adjustment to allow for the variation of the momentum of the machine under different loads.

In Fig. 15, *A* and *B* are the two loose pulleys of the driven shaft, and *C* and *D* the fixed pulleys. *E* is a grooved cam rotating about *J*, and having two roller followers *F* and *G*. *H* is a link driven to and fro by a tripping device attached to the planer bed. *L*, the shifter-arm for the smaller pulleys, is a crank rotated by the follower *F* about *M* as a center. In a similar way, the crank *N* rotates about *O*. The pivots *J*, *M* and *O* are carried on a plate made fast to the planer and not shown in the drawing. The portions of the cam to the left of *F* and to the right of *G* are arcs of circles with *J* as a center, and there-

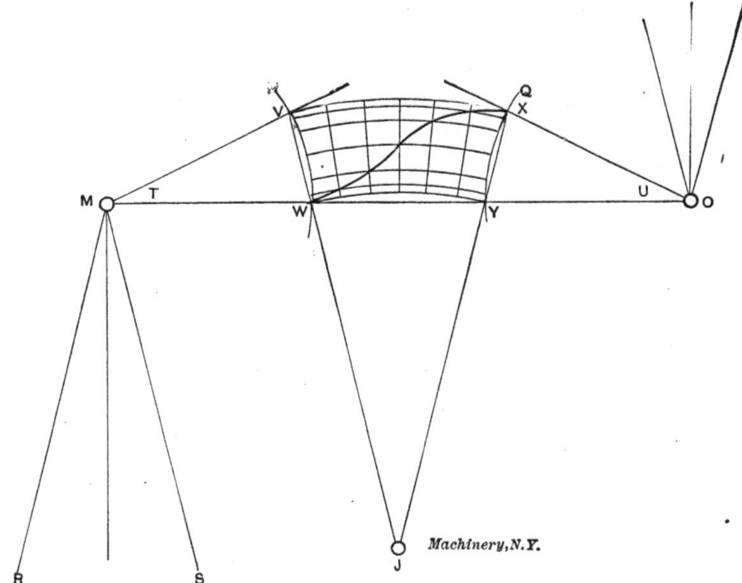

Fig. 16. Lay-out of Cam Curve for Cam in Fig. 15

fore, while either of the followers is traveling through these arcs there will be no movement of the shifter-arms. The throw of either of the arms is occasioned by its follower traversing the irregular path between *F* and *G*.

Imagine the link *H* drawn downwards. The cam then rotates towards the right about the center *J*. The follower *F* is held fixed in its position by the arc of the cam to its left, and therefore the shifter-arm *L* remains stationary. The path of the follower, *G*, however, is through the irregular part of the cam between *F* and *G*, which causes it to rotate about *O* as a center, thereby shifting the arm *N* from the loose pulley *B* to the fixed one *D*. If the link *H* is operated in the reverse direction to that imagined above, the shifter-arm *L* will then become the active member, and the shifter-arm *N* will remain inoperative.

A method for determining the irregular path of the centers of the followers *F* and *G* is shown in Fig. 16. First locate the points *M* and *O* from Fig. 15, and draw the circular arcs *P* and *Q*, the paths of the centers of the followers *F* and *G*. Then draw *R* and *S*, the extreme positions of the center line of the shifter-arm *L*. Make angles *T* and *U* equal to the angle formed by the lines *R* and *S*. Divide the line through *V W* into six parts proportional to 1, 3, 5, 5, 3, 1, and through the points of division draw arcs with *J* as a center. Divide *V X* into six equal parts, through which draw radial lines. The successive intersections of the circular arcs and the radial lines determine the paths of the followers *F* and *G*, as *W X*. Lines drawn tangent to successive positions of a follower along the line *W X* will be the outline of the cam-slot at its irregular part.

The slot *Z*, Fig. 15, permits adjustment of the link as called for in the conditions of the problem. The center of the opening for the belt in the shifter-arm *L* is placed nearer to the center line of the shaft to allow for the angularity of the cross belt.

Laying Out an Intermittent Motion Cam with Pivoted Follower*

The cam to be laid out is shown in Fig. 17. It turns toward the left and moves a 1-inch roller *A* which controls the lever *B* swinging on the stud *C*. The cam is to be keyed to a shaft, together with several other cams, in all of which the keyway is at the beginning of the cycle. The requirements which follow are selected to illustrate as simply as may be the method employed. The head of the lever *B*, which is 12⅜ inches long, is to remain at rest until the cam has turned 150 degrees from the zero point or beginning of the cycle; it is then to advance 1½ inch in 43 degrees; then it will dwell for 35 degrees more, and, finally, retreat 1½ inch in 92 degrees, after which it will dwell for the remainder of the cycle. In Fig. 17 it is seen that the roller *A* is located at one-third of the distance from the pivot of the lever to its head. Hence a movement of one-half inch is required of the roller in the cam to move the lever head 1½ inch.

We will now begin the lay-out. Draw first the circumference of the cam; its diameter we will make 10 inches. With the keyway on the vertical diameter, draw a line through its center. With this line as zero, divide the circumference into 30-degree sections, as shown, and number them. Now draw the circle *D* with a radius of 4 3/16 inches, to show the extreme outer position of the center of the roller, and the circle *E* with a radius of 3 11/16 inches, to show the extreme inner position of the center of the roller. Next, with the center of the cam as its center, draw the circle *F*, so that it will pass through the center of the stud *C*. Beginning with the center of the stud *C* as zero, divide this circle into sections and number them, as shown, for each 60 degrees. Such further sub-divisions as may be needed later may be made when required.

Proceed now with care to place the needle of a pair of good compasses in the center of the roller *A*, and adjust them so that the pencil

* Herbert C. Barnes, MACHINERY, October, 1908.

point will pass through the center of the stud *C*. We will call this radius *R*. Now having in mind the requirements stated above, one being that the cam should turn 150 degrees from its zero before the roller moves, place the end of the compasses at 150 degrees on the circle *D*. Holding the needle here, with the radius *R* draw an arc intersecting the stud circle *F* at the point *G*. It is seen that the point of intersection is at 60 degrees on the circle *F*. Now place the needle point 43 degrees further along on the stud circle, or at 103 degrees, and with the radius *R* draw an arc intersecting the circle *E* at the point *H*. The point *H* marks the halt of the advance of the roller,

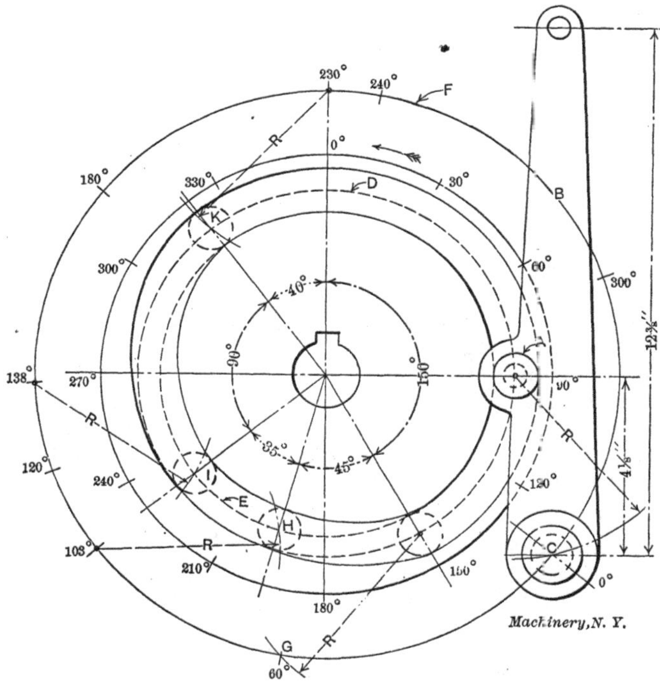

Fig. 17. Lay-out of Intermittent Motion Cam with Pivoted Follower

and the beginning of its dwell. Now move the needle 35 degrees further along the stud circle to 138 degrees, and with the radius *R* draw another arc intersecting the circle *E* at the point *I*. This point marks the end of the dwell and the beginning of the retreat. Now move the needle 92 degrees further along the stud circle to 230 degrees and with the radius *R* draw an arc intersecting the circle *D* at the point *K*. This point marks the end of the retreat and the beginning of the dwell for the remainder of the cycle.

The points *H*, *I* and *K* being marked, draw radii through them extending to the circumference of the cam circle. Knowing that the roller begins to advance at 150 degrees on the cam, the advance is seen to continue for 45 degrees. The roller then dwells for 35 degrees and retreats in 90 degrees, after which it dwells until the next advance begins. It is proper that these figures do not agree with the figures for the lever movement stated above. Barring possible slight errors in the lay-out, they are correct for the cam.

The radius of the inner wall of the raceway or groove is, of course, ½ inch less than that of the path of the cam center. Hence the radius of the inner wall of the outer dwell is 3 11/16 inches, and that of the inner dwell is 3 3/16 inches. This inner wall is the counterpart of the master cam which will be used for cutting the cam groove.

CHAPTER II

CAM CURVES*

When the curve of a cam is not determined by a given definite motion of the follower, and the condition presented to the designer is simply to make the follower move through a given distance during a given angle of motion of the cam-shaft, the ease and silence with which the cam works depends upon the character of curve used in laying out the advance and return. The uniform motion curve, the simplest of all curves to lay out, is a hard-working curve, and one that cannot be run at any great speed without a perceptible shock at the beginning and end of the stroke.

Uniform Motion Curve

The uniform motion curve would be represented in a diagram by the diagonal of the rectangle of which the base represents the angle of motion, and the altitude, the stroke of the cam, as shown by the full lines in Fig. 18. However, should the nature of the design demand a uniform motion for a given part of the revolution of the cam-shaft, the shock at beginning and end of stroke may be modified by increasing both the angle of motion and the stroke, and, in the diagram, filling in arcs of circles as shown by the dotted lines in Fig. 18. The amount of curvature at the ends of the stroke is dependent upon the amount it is possible to increase the angle of motion, and the centers of the arcs are determined by drawing perpendiculars to X Y as shown in Fig. 18. It will be noticed that the uniform motion has been maintained for the original angle, the modifications at the ends causing the increase of angle of motion and of stroke, the rectangle formed by these two being shown by dotted lines. Even with these modifications the cam is still apt to work hard, especially if the angle of motion is small.

* MACHINERY, April, 1907; July, 1907, and February, 1908.

Harmonic Motion Curve

The crank or harmonic motion curve works much more easily than the uniform curve, and a cam laid out with this motion may be run at a high speed without much shock or noise. To draw a diagram of this curve, draw a semi-circle having a diameter equal to the stroke of the cam, and divide this semi-circle and the line representing the angle of motion into the same number of equal parts. The intersection of lines drawn from these divisions will give points on the curve. Fig. 19 shows the harmonic curve and the manner in which it is obtained.

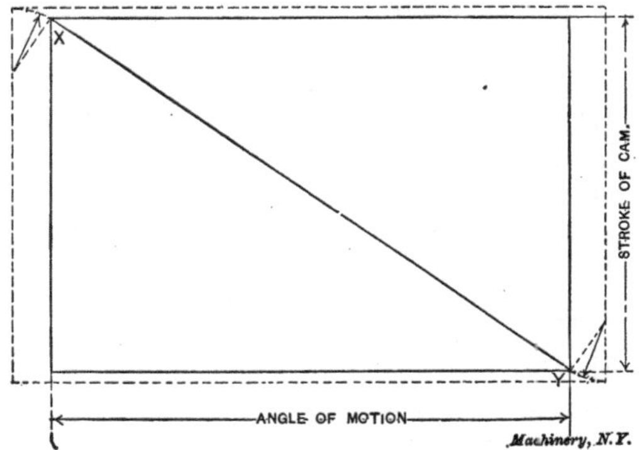

Fig. 18. Uniform Motion Curve

Gravity Curve

Probably the easiest working cam curve is the one known as the gravity curve. This curve has a constant acceleration or retardation bearing the same ratio to the speed as the acceleration or retardation produced by gravity; hence its name. A body falling from rest will pass through about sixteen feet in one second (more accurately 16.08 feet). During the next second the body will increase its velocity by about thirty-two feet making the distance covered during the second

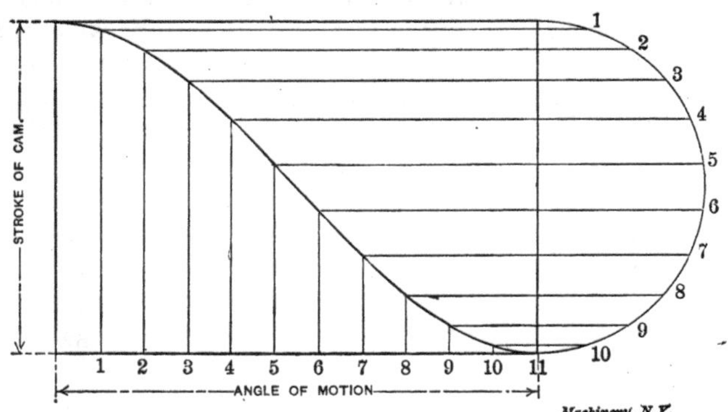

Fig. 19. Crank or Harmonic Motion Curve

second forty-eight feet; during each succeeding second the body will gain in velocity thirty-two feet. Using sixteen feet as a unit of measurement, it will be seen that a body would travel through units 1, 3, 5, 7, 9, etc., during successive seconds or units of time. To apply this motion to the cam curve, we might divide the angle of motion into a given number of equal parts and, using the units given above, we may increase the velocity to a given maximum and then, retarding with the same ratio, bring the follower again to rest at the other end of the stroke. In the diagram, Fig. 20, the line representing the angle of motion is divided into eleven equal parts which necessitates

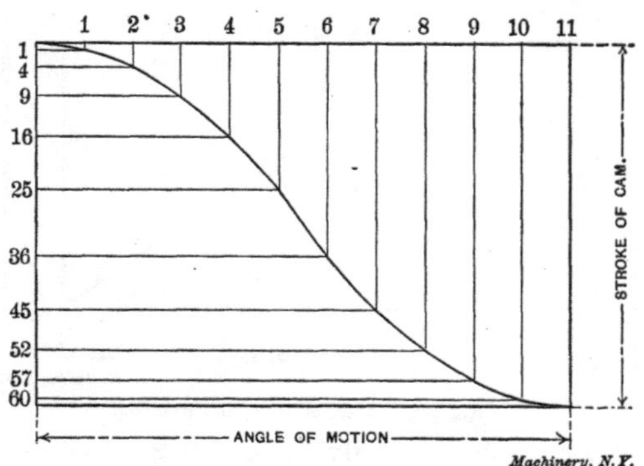

Fig. 20. Gravity Motion Curve

eleven divisions on the line representing the stroke of the cam. If the motion for the first part of the stroke is to have a constant acceleration, as referred to above, the distance traversed by the follower during the first part of the angle of motion would be one unit; in the second part, three units; in the third part, five units, and so on until the maximum velocity had been reached which would be during the

Number of period.	Distance traversed by follower during one period.	Total distance traversed since beginning of motion.
1	1	1
2	3	4
3	5	9
4	7	16
5	9	25
6	11	36
7	9	45
8	7	52
9	5	57
10	3	60
11	1	61

sixth part of the angle of motion when the follower would travel through eleven units of motion. At this point the motion would begin to be retarded by a constant deduction which would cause the follower to move through nine units during the seventh interval of time, seven units during the eighth, five units during the ninth, three units during the tenth, and one unit during the eleventh and last interval. The sum of these units is sixty-one, which will necessitate dividing the line representing the stroke of the cam into sixty-one equal parts of which the first, fourth, ninth, sixteenth, twenty-fifth, thirty-sixth, forty-fifth, fifty-second, fifty-seventh, sixtieth, and sixty-first will be used for determining points on the curve. The combination of the table given and the diagram shown in Fig. 20 will show how the gravity curve may be drawn.

Approximation of Gravity Curve

A very close and satisfactory approximation for the gravity curve, and one that entails less work than the theoretical, is shown in Fig. 21. The method of drawing is similar to the one used for the harmonic motion, excepting that an ellipse takes the place of the semi-circle. It can be seen very readily that the ratio of the major and minor axes will determine the character of the cam curve. To obtain a curve that

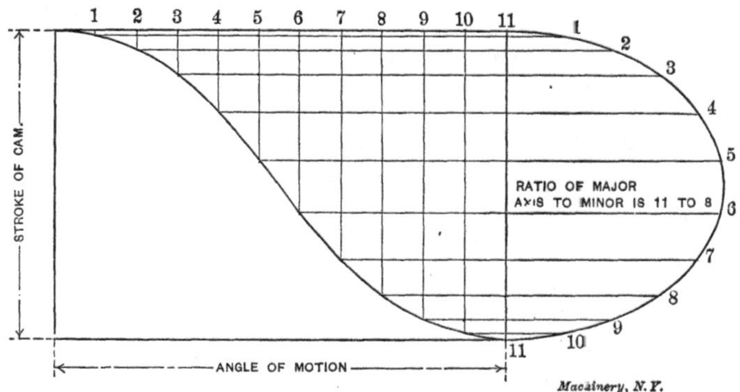

Fig. 21. Approximate Gravity Curve

will approximate the gravity curve, the line representing the stroke of the cam should be used as the minor axis and the ratio of major axis to minor axis should be 1⅜ to 1 or 11 to 8. Dividing the semi-ellipse and line of angle of motion into the same number of equal parts, and projecting, we obtain points on the curve. Fig. 22 is given so that a comparison may be made of the three motions given above when applied to the same cam.

Laying Out Cams for Rapid Motions

As already mentioned in Chapter I, we may consider a cam mechanism as being made up of two elements. As generally constructed, one element is a revolving plate cylinder, cone or sphere, and the other element is a bar or a roller which has some form of reciprocating motion. The revolving piece is usually made the driver, although the mechanism may be made to work in the reverse order. The shape of a cam will depend upon the kind of motion that the follower is required to have. The motion of cams that are used for driving parts of machinery, may be, as we have already seen, one of three kinds, *viz.:*

1. *Uniform motion*, in which the follower is made to pass over equal spaces in equal intervals of time.

2. *Simple harmonic motion*, in which the follower is accelerated from rest to a maximum velocity and then retarded again to a state of rest, following the harmonic cycle.

3. *Uniformly accelerated motion*, in which the follower is accelerated from rest to a maximum velocity and then retarded again to a state of rest, the acceleration being uniform, as, 1 inch per second, 2 inches per second, etc.

To this we may add a fourth kind frequently met with:

4. *Intermittent motion*, periods of motion being interrupted by periods of rest.

In slow-moving machinery it may not be important whether the follower moves with uniform, simple harmonic, or uniformly accelerated motion, but in machines where the cams have a high rotative speed, and the follower a reciprocating motion, as in the case of sewing machines and in some textile machinery, a uniform rate of motion will

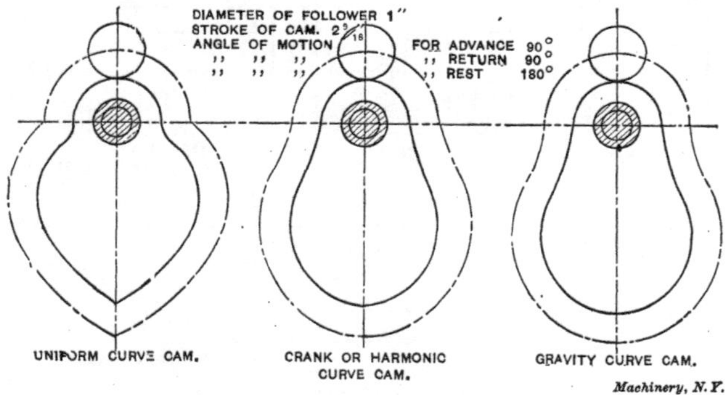

Fig. 22. Comparison between the Different Cam Constructions

be unsatisfactory or impossible. The reason for this is that the follower is impelled from rest to its maximum velocity instantly, and also brought to rest from a maximum velocity instantly. This gives it a sudden jerk at each end of the motion, which is very trying to a machine when the reversals take place rapidly. Cams for high rotative speeds, where the follower has a reciprocating motion, should, therefore, be so designed that the follower will start gradually, attain its maximum speed near the middle of its path, and then gradually come to rest. In other words, the follower should have a uniformly accelerated motion during the first half of its movement, and a uniformly retarded motion during the last half.

In uniformly accelerated motion $S = \frac{1}{2}Pt^2$, where S = the distance passed over, P = the acceleration, and t = the time. This is the same as saying that the distance which the body has passed over at the end of any number of units of time varies as the square of the number of such units. For example, if a body has a uniform acceleration of 2 inches per second, $S = \frac{1}{2} \times (2) \times (1)^2 = 1$ for the first second; $S = \frac{1}{2} \times (2) \times (2)^2 = 4$ for the next second; and so on. This is, as said before, also the law of falling bodies whose motion is not resisted by the air or other medium. Uniformly retarded motion obeys the same law. If time intervals of such a motion be plotted as abscissas and the corresponding space intervals as ordinates, with reference to co-ordinate axes, the resulting curve will be a parabola, and this is the curve that should be used for the outline of cams that are designed for high rotative speeds.

Uniform Motion Cylinder Cam

The cams shown in the following cuts do not necessarily represent any existing forms; they simply illustrate how the principle may be applied to certain shapes of cams and paths of followers. In Fig. 23, lay out on a sheet of paper *A B D C* a line constructed as follows: Bisect *C D* at *M* and divide *C M* into any convenient number of parts, say

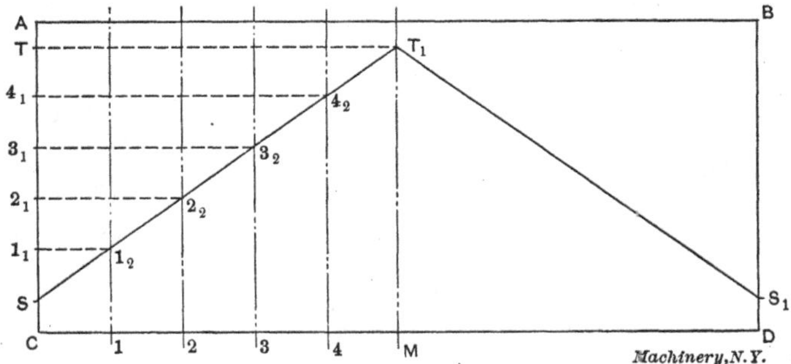

Fig. 23. Development of Uniform Motion Curve

five. Lay off on *C A* any distance *S T*, and divide *S T* into the same number of parts as there are in *C M*. Through the points 1, 2, 3, etc., on *C M*, erect perpendiculars to *C M*, and through the points 1_1, 2_1, 3_1, etc., on *C A*, draw parallels to *C M* intersecting the perpendiculars at points 1_2, 2_2, 3_2, etc. A line $S\,T_1$ drawn through these intersections will be straight. The line $T_1\,S_1$ can be found in the same way. Now if the sheet of paper *A B D C* be wrapped around the outside of a cylinder whose circumference is equal to the distance *C D*, the line $S\,T_1$ will take the position *S T*, Fig. 24, and the line $T_1\,S_1$ will form a similar curve on the reverse side of the cylinder. If this curve be made the center line of a groove, as the cylinder revolves on its axis, the groove will drive a follower up and down, parallel to the elements of the cylinder, with a uniform speed. The follower will start and stop at either end of its motion with a sudden jerk.

Uniformly Accelerated Motion Cylinder Cam

In Fig. 26 let *A B D C* represent the paper as before. Bisect *C M* at 3, and *S T* at 9. Divide *C*3 and 3*M* into any convenient number of parts, say three; then divide *S*9 and 9*T* into the square of three parts, or 9, as shown. Erect perpendiculars to *C M* at the points 1, 2, 3, etc., and draw parallels to *C M* through the points 1, 4, 9, 4_1, and 1_1. Through the points S and T_1 and the intersections 1_1, 2_1, 3_1, $2'_1$ and $1'_1$, draw a smooth curve. This line will be a parabolic curve, reversing at 3_1. The curve $T_1 S_1$ is constructed in the same way. Now wrap the sheet of paper $A\,B\,D\,C$ around a cylinder whose circumference is equal to $C\,D$. The curve will take the position $S\,T_1$, Fig. 27, and the curve $T_1 S_1$ will take a similar position on the reverse side of the cylinder. A groove made with these curves as center lines will drive a follower P up and down through the distance K, as the cylinder is rotated on its axis. The follower will start gradually at S, attain its maximum velocity, and then come gradually to rest again at T_1, the motion being

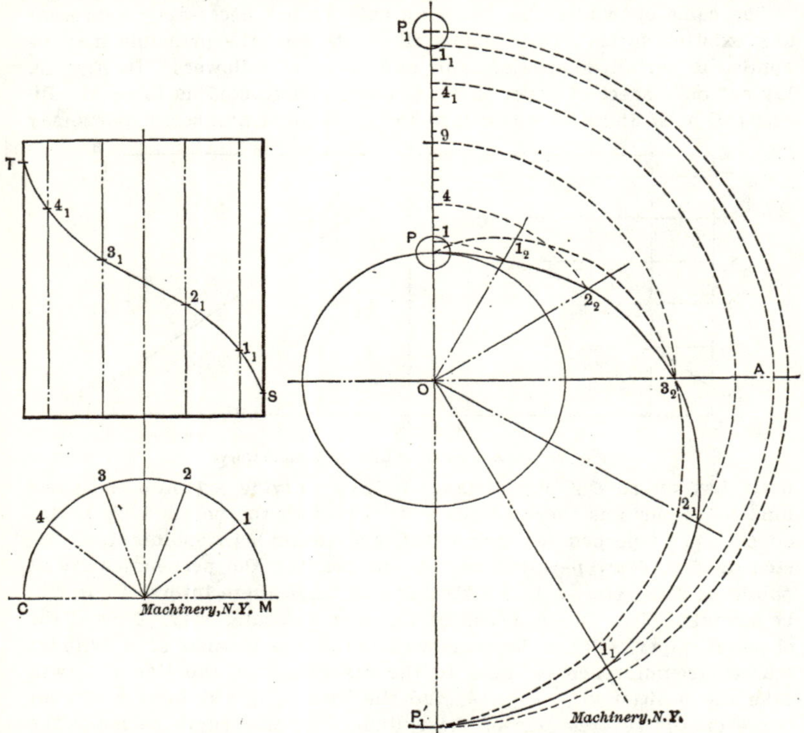

Fig. 24. Uniform Motion Curve scribed on Cylindrical Surface

Fig. 25. Accelerated Motion or Gravity Curve applied to Plate Cam

uniformly accelerated and retarded. The sides of the groove are made parallel to $S\,T_1$, and drawn to suit the diameter of follower P.

Fig. 28 shows the distortion of the curve $S\,T$ when the follower moves in the arc of a circle, with center at some point Q, instead of in a straight line. Points on the new curve are found by setting off from the intersections b_1, d_1, etc., the ordinates $a\,b$ and $c\,d$. The curve $S\,a_1\,c_1\,T$ is then made the center line of a groove which will drive the hinged follower with the same variation in speed attained by the follower in Fig. 27.

Accelerated Motion Plate Cam

Fig. 25 shows how the parabolic curve is applied to a plate cam. The roller follower is supposed to oscillate between P and P_1 as the cam rotates about O. The curve $P3_2P_1'$ corresponds to $S\,T_1$ in Fig. 27, being the center line of the parabolic groove in the face of the plate.

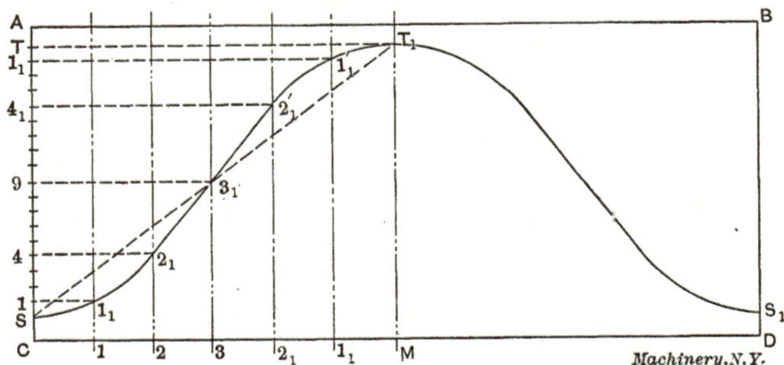

Fig. 26. Development of Uniformly Accelerated Motion Curve

Only one-half of the cam is shown in the figure. Suppose this cam is to rotate 180 degrees, while the follower moves from P to P_1. Draw the base circle with radius $O\,P$, the length of which will depend upon the size of the cam. Draw $O\,A$ perpendicular to $O\,P$, and divide the arc subtended by $P\,O\,A$ into any convenient number of parts, say three. Draw radii $O1_2$, $O2_2$, etc. Divide $P\,P_1$ into two equal parts at 9, and divide $P9$ into the square of three parts, or 9, as shown. With O as a

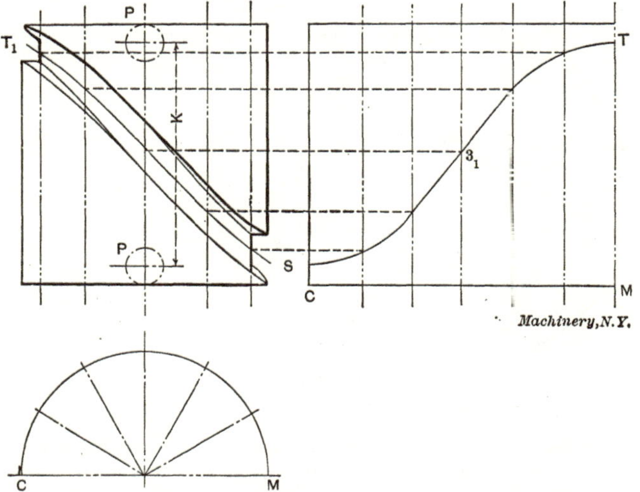

Fig. 27. Transferring Uniformly Accelerated Motion Curve to Cylinder

center, and radius $O1$, find the intersection 1_2. In the same way find the other intersections 2_2, 3_2, etc., and draw a smooth curve through these points. This curve has the same relation to the curve of uniform motion shown dotted, that the parabolic curve has to the straight line in Fig. 26. If a similar curve be laid out on the other side of $P P_1'$, and made the center line of a groove, then the follower P will be pushed up and down mechanically by direct contact. If a curve parallel to $P3_2 P_1'$, and drawn at a distance equal to the radius of the follower away from

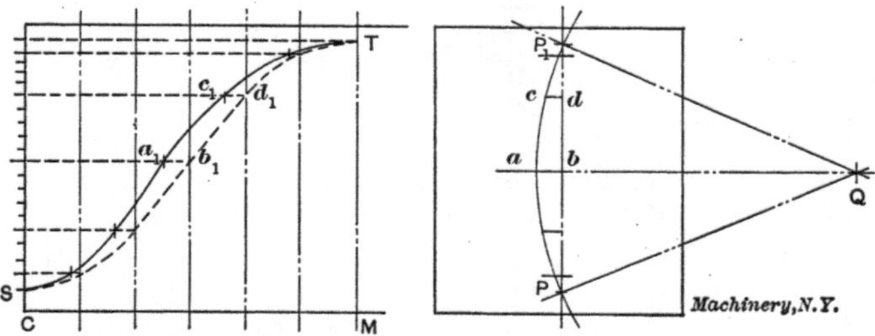

Fig. 28. Accelerated Motion Curve, when Follower moves in the Arc of a Circle

it, on the inside, be made the outline of the cam, then the follower will be pushed up mechanically to P_1, and allowed to fall by its own weight. It will remain in contact with the cam theoretically, because the principle of uniformly accelerated motion is the same as that of a falling body. In practice, however, the friction and the inertia of the connected

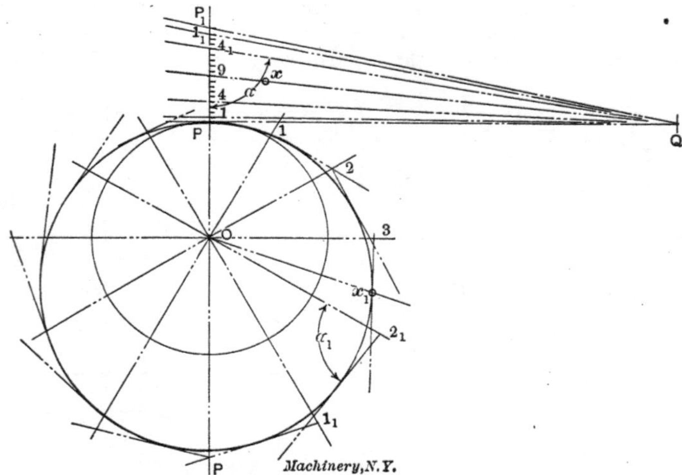

Fig. 29. Plate Cam for Bar Follower

parts would probably prevent the follower from remaining in contact with the cam on its return motion if the oscillations were rapid.

Fig. 30 shows the parabolic cam constructed for a follower which moves in any curved path. The construction is the same as in Fig. 25 except that points on the curve are located on radial lines Oa_1, Ob_1, etc., offset from the first radii by the distances $2a_1 = 4a$, $3b_1 = 9b$, and so on.

Plate Cam with Bar Follower

When a plate cam is to be laid out to drive a bar follower through a certain cycle of operations, the construction is more complicated. The base circle is divided as in the previous case into any convenient number of parts, and the square of the number of such parts laid out from P to 9 and from 9 to P_1, Fig. 29. If the bar is to oscillate about Q as a center, it will take the positions $Q1$, $Q4$, $Q9$, etc., as the radii $O1$, $O2$, $O3$, etc., come to the position OP. The intersections 1, 2, 3, and so on, are found just the same as in the previous cases. Now instead of drawing the curve for the cam outline through these points, straight lines which represent the edge of the follower must be drawn

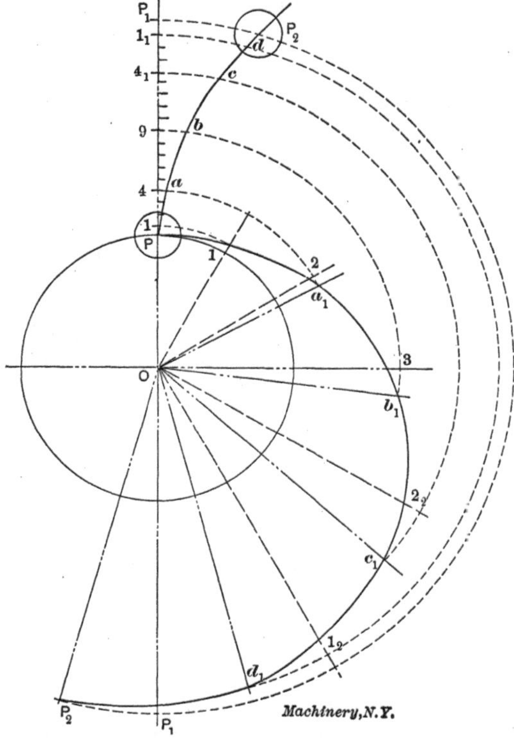

Fig. 30. Accelerated Motion Curve applied to Plate Cam, with Follower moving along a Curve

through the points making the same angle with a given radius as the follower makes with OP when the radius in question is in the position OP. For example, angle a equals angle a_1. Now the cam outline is a smooth curve drawn tangent to these straight lines. If the bar follower, instead of being centered at Q, moves up and down parallel to its first position, then all these angles are right angles. If the face of the bar is curved, then the cam outline must be drawn tangent to the curves after they have been properly located with respect to their several radii.

In drawing cams like Fig. 29, the proper relation between the diameter of the base circle and the distance PP_1 must be assumed. If the base circle is too small, the cam outline will not be tangent to the edge of the follower in all positions, and the latter will not have uniformly accelerated and retarded motion. There is a rolling and sliding contact between the cam and its follower in the case of Fig. 29. The rolling action tends to carry the point of contact outward to the right of *OP*, during the upward motion, and to bring it back towards *OP* during the downward motion. The point of contact x does not necessarily occur when Ox_1, is perpendicular to Qx.

Effect of Changing Location of Cam Roller

When the line of motion of a follower passes through the center of rotation of the cam and the angle of the curve causes it to work hard,

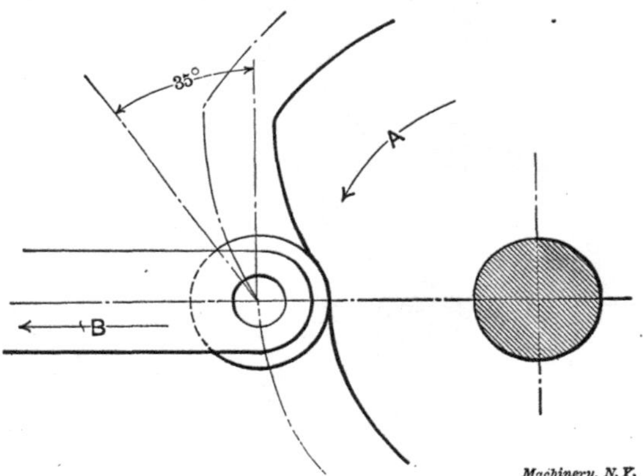

Fig. 31. Cam Roller on Center Line of Cam

the curve may be modified, and the same motion of follower obtained by placing the follower with its line of action parallel to its original position and not passing through the center of the cam. A condition may be assumed, as shown in Fig. 31.

Here we have a cam, rotating in the direction indicated by the arrow *A*, whose duty it is to move the follower ¾ inch in the direction indicated by the arrow *B* during a 30-degree angle of motion of the cam-shaft. The angle of the cam as presented to the follower at the beginning of the stroke would be 35 degrees, as determined by the tangent to the curve of the centers, as indicated on the drawing. After the follower had moved one-third of its distance, the angle presented would be 32 degrees, and when two-thirds of the travel had been made, the angle of the curve would be about 30 degrees. The angles given are for a curve which would give a uniform motion to the follower. Should the cam curve work hard at the required speed we would naturally make the cam of greater diameter, if possible, which would reduce the angle of the cam, as shown by the difference in the angles presented in Fig. 31, as we go out from the center of rotation. The design of the machine, however, might make this change impossible. If it was simply necessary to get the follower from the position shown to a point ¾ inch distant in a 30-degree movement of the cam-shaft, without regard to its motion, a harmonic or gravity curve might be used which would cause the cam to work easier. However, this would be impossible should our design require a uniform, or some other equally hard motion. A third way in which the angle of the curve might be decreased would be to make the angle of motion of the cam-shaft greater. This, too, might be made impossible by the limitations of our design.

Another way, and one not commonly used, consists in changing the location of the cam roller. In Fig. 32 all conditions are the same as

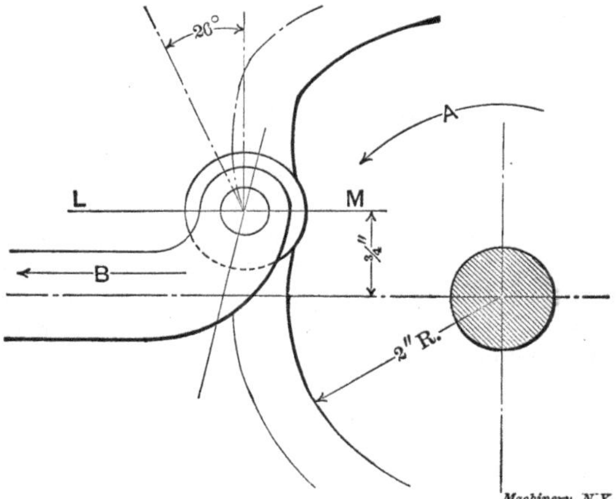

Fig. 32. Cam Roller placed above Center Line of Cam

in Fig. 31, except the roller has been placed ¾ inch above the line passing through the center of the cam. The center of the roller will now pass along the line *L M*, or parallel to the line of motion in Fig. 31. The angle of the curve presented to the roller in this case is 26 degrees, much less than the angle presented in Fig. 31, and the angle decreases as the roller moves away from the center of rotation. The advantage that may be gained by moving the cam roller may be readily seen by comparing the results given above. There is, of course, a limit to the distance the roller may be changed, for if placed too far away from the center line, the thrust in the direction at right angles to the direction of motion of the follower would be so great as to offset the advantage gained.

Even without the aid of an illustration it may be seen that to place the cam roller on the other side of the center would cause the angle of the cam curve to increase, thus making conditions worse. The offset of the roller should be in the direction opposed to the direction of motion of the cam.

CHAPTER III

NOTES ON CAM DESIGN AND CAM CUTTING*

It is strange that the processes and methods of cam cutting have not been improved more rapidly than they have. Twenty-five years ago, cams and gears were on about an equal footing; that is to say, most of both were cast to as nearly the proper shape as possible, after which the working surfaces or teeth were smoothed up with a file, and then the holes and hubs were finished in the usual manner. Some cams of both plate and barrel forms were cut, with suitable attachments, in the same machine the gears were cut in. This was an old hand indexing machine, with an automatic feed composed of a weight hung on the pilot wheel. Since that time gear cutting machinery has been wonderfully developed. All sorts of styles and arrangements are on the market, meeting every demand, from that for a general purpose machine to highly specialized forms. When it comes to cam cutting machinery, however, while machinery builders have special tools for their own work, so far as the writer is aware, there is no tool regularly on the market for cutting cams. The cam has thus fallen behind the gear in the process of development. Machine designers and machine users are liable to be a little suspicious of cams, anyway. Considerable trouble is often taken to avoid the necessity for using them. This is due, however, as much to faulty design and faulty construction as to any inherent objections to this form of mechanical movement. It is here proposed to call attention to some of the points to be considered in designing and producing satisfactory cams, with the thought of thereby doing something to justify a more extensive use of them.

Faults in the Design of Cams

We have all seen cams that were the cause of a good deal of profanity, in which the trouble could be traced to the designer or machinist, who laid out the curves on what might be termed "schedule time"; that is to say, he simply made sure of his starting and stopping points, neglecting all intermediate points so long as the movement got there and got back on time. This, he thought, would be all that was necessary, not taking into account the shock and jar caused by the sudden starting and stopping of heavy slides, levers, etc., at even moderate speeds. The temptation to do this is always strong, especially in the case of barrel cams, where it is so much easier to use the milling machine (gearing it up for a spiral to meet the schedule requirements), than it would be to lay out and form a curve with a gradual starting of the motion and a gradual stopping. There is nothing worse for the life of a machine than to have it operated by cams cut by this "sched-

* MACHINERY, August, 1907.

ule" method. Another point to consider is that of taking advantage of all the time there is for any given movement. The period or periods of rest should be cut down to the last degree, so as to have the angularity of the rise as small as possible. Careful work at the drawing-board will make a big difference with the satisfactory action of cams in these two respects. Still another bad practice, which has perhaps tended to throw the use of cams into disfavor, is that of making them in two or more parts, with the idea of having the working surfaces adjustable. After they have been wedged out, or shimmed up, or ground off a few times, a more proper name for them would be "bumpers" rather than "cams." Except in rare cases, there is no more use or excuse for adjustable cams than for adjustable gears, as there are other and better means of making adjustments when these are necessary. Cams are not very expensive as compared with gears, and they can be duplicated with greater accuracy than most machine parts. Especially is this the case if roughing and finishing mills are used in forming them, as the finishing mill will retain its cutting edge and size for a great number of cams, if it runs true with the spindle in the first place.

Cam Rolls and Roll Studs

A few words might be said with relation to the design and construction of cam rolls and the studs for them, since the successful working of a cam depends to a considerable degree on this matter. The design of the roll and its stud should be such that the work it has to do, the speed at which it runs, and the bearing area on the stud, should be the factors determining its size, rather than the simple fact that there is a milling cutter in the tool-room of a certain diameter. It is equally important that the roll and stud should be ground all over after hardening. The end of the roll should also be cut back for 1/64th of an inch or so on the sides for some distance from the outside diameter, so as to avoid undue friction against the collar of the stud, or the part it is mounted in. On account of the warping that takes place in hardening, rolls that are not ground inside and out have a habit of stopping frequently under load, until in time flat spots are worn on the face; then the working surface of the cam will begin to wear or rough up. Roll studs that are the slightest degree out of parallel to the working surface of the cam will also cause some trouble, but no amount of grinding will help this case. The same trouble occurs on barrel cams if the milling cutter is set above or below the center of the cam when cutting it. The roll will then bear at one end only at the most important time, when the throw takes place. A conical roll is the proper thing for this style of cam. There is a lot of end pressure to a roll of this type, however, which must be taken care of by thrust collars on the stud; or, better still, a ball race may be scored in the collar and the large end of the roll, so as to provide for a ball thrust bearing. This end pressure will reduce the side pressure on the stud to quite an extent, nevertheless, so the latter may be made slightly shorter or smaller in diameter than when a parallel roll is used.

Cutting Cams of Uniform Lead in the Miller

When it comes to the cutting of cams, the shop man naturally turns to the milling machine. Many manufacturers of milling machines make attachments which may be used for cutting cams with formers. None, however, is provided with anything except hand feed. Another, and the greatest, objection to them is that if there is much work to be done, one of the most expensive machines in the shop is tied up, and there are few shops that have a surplus of this brand of machine tools. For an occasional or an experimental job, however, there is nothing better than the milling machine. As has been before remarked, curves with easy starting and stopping movements cannot be cut

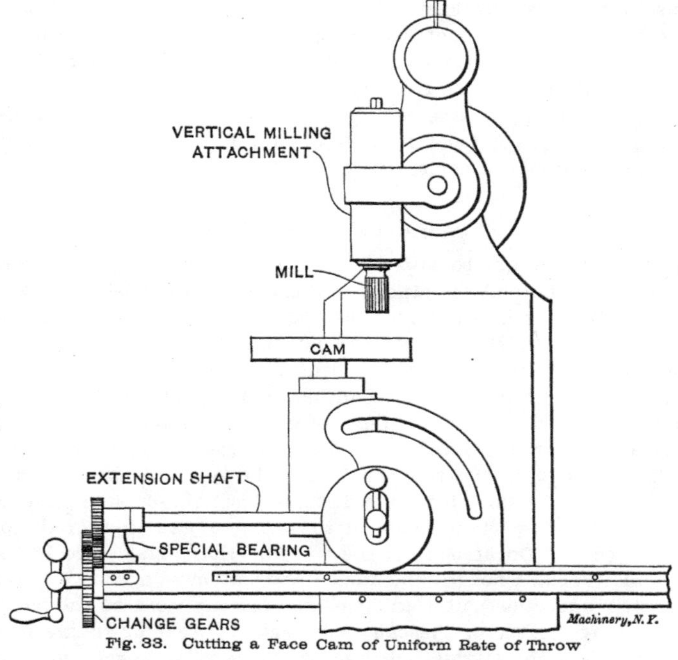

Fig. 33. Cutting a Face Cam of Uniform Rate of Throw

without formers on it, or on any other machine for that matter; but cams which require a constant rise, such as the feed cams of some machines, may be cut on it without the use of formers. With barrel cams the method is obvious, it only being necessary to gear the spiral head with the lead-screw to get the required lead, and then cut a groove of this pitch in the body with an end mill of the same diameter as the roll.

For cutting plate cams for the same kind of motion, the arrangement shown in Fig. 33 may be used, if the machine happens to have a vertical spindle milling attachment and a spiral head. All that it is necessary to provide in addition is the extension shaft shown, and the special bearing or bracket for supporting it. These parts are used to bring the spiral head to the center of the table. The shaft is bored out at one end to fit the stud of the spiral head (called the worm gear stud in the tables); the other is turned and keyed to fit the change gears. The cams may be held in the regular chuck, or on a face-plate fitted to the head. Small ones may be held on an arbor fitted to the spindle, with large collars to hold them firmly, clamped with a nut and washer, or by an expansion bushing in the case of large holes. If they have keyways in them, and more than one or two are to be made, it will be well to fit a key in the arbor to help locate them. It is necessary to set the mill central with the spiral head to obtain correct results, as the spiral will vary if this is not done. Advantage may sometimes be taken of this when, with the regular change gears, there is no spiral of the exact pitch required, in which case the desired rise

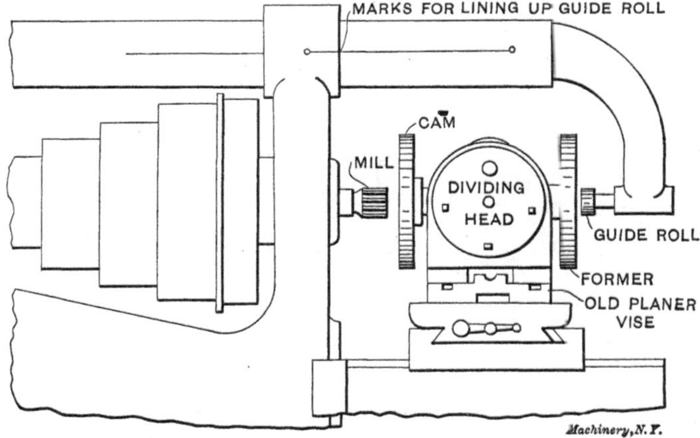

Fig. 34. Inexpensive Fixture for Milling Plate Cams to Match a Former

can be obtained by setting the head off center. This, however, will not give a uniform spiral, as the pitch will keep increasing as it leaves the center of the cam. As cam drawings are generally laid out or divided in degrees, it will be found convenient to divide the cam blank by the same method, while held in the spiral head. To do this, we may revolve the index crank through two holes in the 18-hole circle or three holes in the 27-hole circle, as many times as are necessary, each of these divisions giving exactly one degree.

Milling Machine Attachments for Cutting Cams with a Former

Examples of attachments rigged up to suit special requirements are shown in the cuts Figs. 34 and 35. To a shop with a rather limited equipment, an order came in for a lot of eight machines, which required seven cams each, most of which were of the plate type. As this class of work was new to the shop, there were no facilities for this part of the job; as usual, it was decided to do the work on the milling machine.

An old planer vise was scraped up and refitted so as to have the movable jaw a nice sliding fit—the screw having been removed, of course. To this jaw was fitted and bolted the spiral head of the miller, in such a way that its spindle could be placed either at right angles, or parallel to the cutter, as the case required for barrel or plate cams. An arbor was made, long enough to pass through the head, carrying the former on the back end and the cam blank on the front end. A nut threaded onto the back end held the former against the end of the spindle, so there was no danger of the arbors rattling loose, no matter how badly the work and tool chattered.

For plate cams, as shown in Fig. 34, the former was made the opposite hand to that of the cam required. The overhanging arm had a center line marked on it as shown, which was matched with one on the frame so as to locate the arbor support central with the spindle. In the place of the arbor-supporting center there was fitted a stud

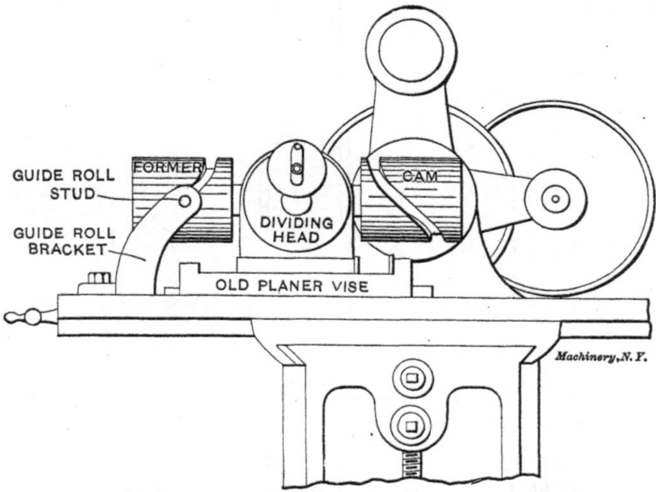

Fig. 35. Cutting a Cylindrical Cam with the Attachment shown in Fig. 34

with a roller of the same diameter as the cutter. The arm was held securely by the regular milling machine braces, which are not shown in the cut. The method of operation is obvious. The spiral head with its attached work and former was revolved, slowly, by hand. The action of the roller, held by the overhanging arm in the groove of the former, caused the head and work to slide back and forth on the ways of the planer vise, giving the proper movement between the work and the cutter to produce the desired contour of cam. The table was locked on the saddle.

For barrel cams, the attachment was rearranged as shown in Fig. 35. The former roller was held firmly in a bracket bolted to the table of the machine. As the roller is on the opposite side of the milling cutter, the former and work are set 180 degrees apart on the work arbor, otherwise they are alike. The head is relocated on the movable vise jaw to bring the axis of its spindle at right angles to the axis of the cutter, as shown. The reader will easily make out the other details from the engraving.

Both arrangements cut good cams, considering that the first cost of the whole outfit was very little. As the formers were made accurately to drawing, the cams gave good satisfaction at fairly high speeds, but the device had the disadvantage of tying up a machine which had plenty of work waiting for it; besides, it was a tedious job to feed the index crank by hand all day long, especially when working on steel cams. For these reasons, when a duplicate order came in, a few weeks later, it was considered best to try the plan of cutting the plate cams on an old lathe, thus providing the advantage of an automatic feed, and relieving the miller of some of its work as well.

A Face Cam Cutting Attachment for the Lathe

A lathe cam cutting attachment is shown in Fig. 36. While not new in principle, it differs somewhat from the other makeshifts described.

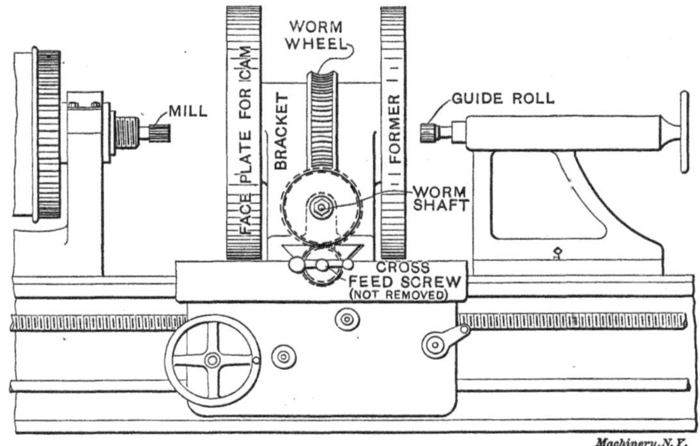

Fig. 36. Attachment with Power Feed for Cutting Face Cams

For this arrangement, the tool slide was removed from the machine and replaced with the bracket casting shown. This was fitted and gibbed to the tool-rest slide, and had its spindle bored and sides faced with a boring bar on the lathe centers. To the bracket was then fitted the cam face-plate and spindle, cast in one piece and finished all over, with the back or small end threaded to fit the former. Keyed to this spindle was a worm-gear of cast iron. In this case the worm-gear had 82 teeth. Meshing with this gear was a worm having 9/16 inch hole, and with a key having a sliding fit in the worm shaft. Bearings were provided for the worm shaft at front and back. The front support for the worm shaft was cast onto the bracket, and finished with it to fit the tool-rest slide, after which it was sawed off and fastened at the front of the carriage by the gib screw, as shown. This is the same practice as is commonly followed in making the clamp for the threading stop on the cross slide. To the outer end of the worm shaft was keyed a gear, meshing with another fitted and keyed to the front end of the cross feed screw next to the handle. The quill was cut off to make room for it. The cross feed nut was removed entirely, of course.

It will be seen that this arrangement, while having the general features of that shown in Fig. 34, provided the advantage of making use of a less costly and less over-worked machine, and allowed the use of a power feed as well, since the gearing provided for connection with the power cross feed in the apron. This gave a feed fine enough for small cams, but on large ones it was necessary to run the feed belt from the feed shaft cone to the hub of the large intermediate gear of the screw-cutting train, this being in mesh with the spindle gear. The lead-screw was removed so as not to interfere with the belt. With regular changes this gave a wide range of feeds.

The cams and formers were held to their respective face-plates by bolts. All the formers were of the positive follower type having a groove for the guiding of the roller. No weight or other means is then required for the followers to hold them to their work.

CHAPTER IV

CUTTING MASTER CAMS*

Common Method of Making Master Cams

Assuming that the master cam has been properly machined and roughed down, we will consider briefly the generally used method of finishing it. This method comprises mounting the master cam in the dividing head of a universal milling machine, and gearing the head with the feed-screw of the table so that the table will advance in proper ratio with the turning of the work in the dividing head. In Fig. 37 a master cam is mounted as above described, and held against a cutter in the vertical spindle milling attachment on a milling machine. This cutter is of the same diameter as the roll which will be used with the cam. **The following description refers specifically to the cutting of the master cam for the cam shown in Chapter I, Fig. 17.**

The process is as follows: Feed the work against the cutter until

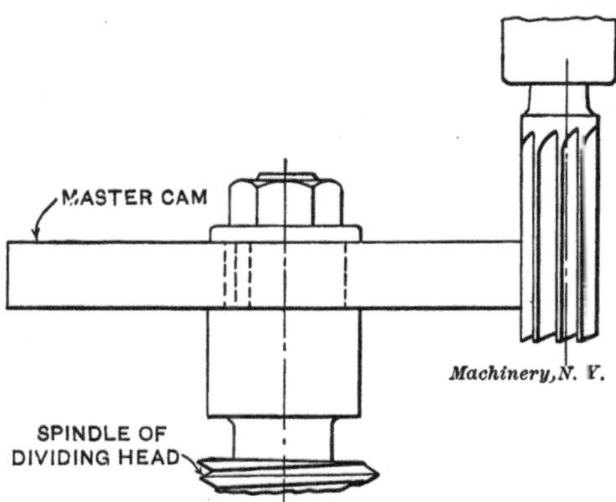

Fig. 37. Common Method of Milling Master Cams

the cutter is 3 11/16 inches from the center of the master cam. Now, with the key-slot of the master cam which is the "zero" of the cam, directly in line with the cutter, turn the work 150 degrees. This finishes a part of the outer dwell of the cam. The next operation is to feed the work against the cutter ½ inch while the dividing head turns 45 degrees. Since 45 degrees is ⅛ of 360 degrees, or one turn, we want gears which will turn the work ⅛ of a revolution while the table advances ½ inch. This is equal to one turn of the work while the table advances 4 inches. The gears on a feed-screw with four threads per inch, and 40-tooth worm-gear in the dividing head are:

Gear on worm 36,	Second gear on stud 28,
Gear on worm 36,	Gear on screw 70.

* MACHINERY, October, 1908.

Having connected these gears with care, feed the work against the cutter 0.500 inch. The gears will at the same time turn the work 45 degrees. This will give the advance of the cam. Now, with the table clamped where it is, turn the work 35 degrees further. This will give the inner dwell of the cam. Now change the gears so that the work will turn 90 degrees while the table is backed away ½ inch. This may be done by removing the first gear on the stud with 36 teeth and replacing it with a 72-tooth gear. Having done this with care to avoid disturbing the work during the change, back the work away from the cutter 0.500 inch. The gears will have turned the work 90 degrees more, the intermediate having been properly adjusted. This will give

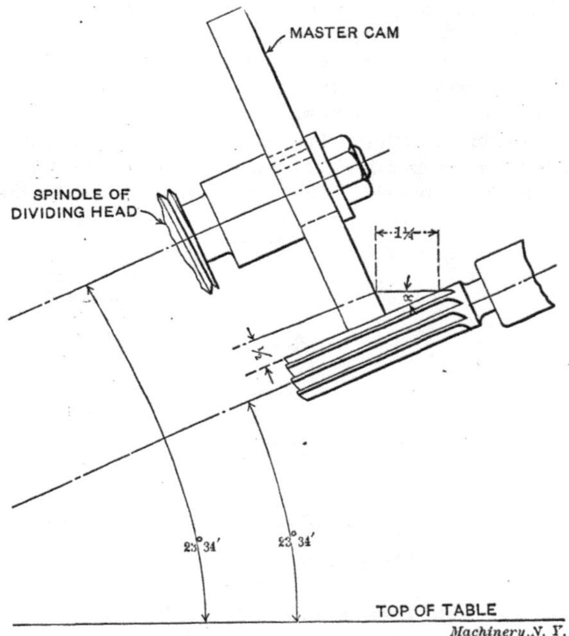

Fig. 38. Improved Method of Milling Master Cams

the retreat of the cam. Now, with the table clamped where it is, turn the work until the cutter reaches the part already finished.

The method which has just been described, is very convenient when the change gears will give the combinations that are necessary, but it often happens that the desired combination cannot be made with even an approach to accuracy. This difficulty may be overcome, however, by a method which is not in general use, but by which any desired result may be obtained.

Improved Method for Producing Master Cams

For convenience we will suppose that the master cam could not be cut with the gears named or with any others, in the vertical position.

We will proceed as follows: Mount the roughed-out master cam as before in the dividing head, and place a 1-inch end mill in the vertical milling attachment, but, instead of setting them in a vertical position, incline each at an angle of 23 degrees 34 minutes, as shown in Fig. 38. The reason for this will appear later.

By inspection we see that if the work be fed against the cutter, Fig. 38, the cutter will enter the work and approach the mandrel. We also see that if the angle of inclination be increased or reduced, the

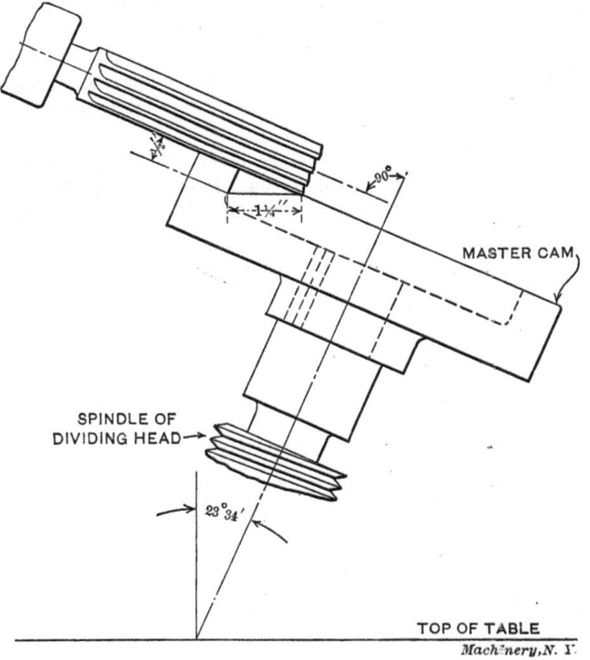

Fig. 39. Milling a Master Cam for a Drum Cam

rate with which the cutter approaches the mandrel will vary likewise. A convenient combination of gears to use in this case is one which will turn the work 360 degrees while the table advances 10 inches. This result may be obtained by using four 36-tooth gears to turn the work.

Having milled the master cam for the first 150 degrees to a radius of 3 11/16 inch as mentioned, we must find the correct distance to feed the table forward in order to make the cutter approach the mandrel ½ inch while the work turns 45 degrees. The computation is done as follows: Forty-five degrees is ⅛ of 360 degrees. Since the table is geared to advance 10 inches while the work turns 360 degrees, the table will advance ⅛ of 10 inches while the work turns 45 degrees. Thus the advance is 1¼ inch to the 45-degree turn of the work. By inspection we see that in Fig. 38 the cutter and the work-face form two sides in a right-angled triangle with a hypothenuse of ¼ inch and one side of ½ inch. By solving, we find the angle a to be 23 degrees 34 minutes, as before mentioned. Having now properly connected the gears to mill the advance on the cam, feed the table ahead 1.250 inch. As just stated, this will make the cutter approach the mandrel ½ inch while the gears will have turned the work 45 degrees. Now with the table clamped where it is, turn the work 35 degrees more. We are then ready to begin the retreat of the cam. We must arrange gears which will turn the work 90 degrees while the table is backed 1¼ inch. By removing the 36-tooth gear from the screw and replacing it with a 72-tooth gear, we get this result. Carefully make the change so as not to disturb the work, and then back the table 1.250 inch. The gears will have turned the work 90 degrees further. Now, with the table clamped where it is, turn the work until the master cam is completed.

This system for making cams may be used only where uniform movements are required. While we have used it to entirely finish a master plate cam, any part of any cam requiring uniform motion may be

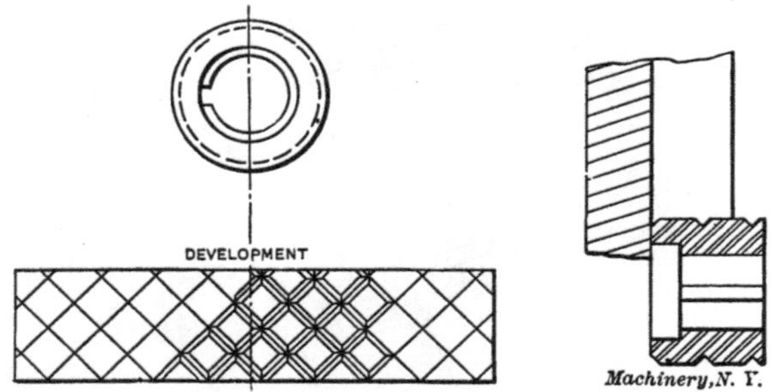

Fig. 40. Special Finishing Cutter for Cam Grooves

milled in this way with a degree of accuracy not readily obtained in any other way. In fact, the work should be as true as the machine on which it is done. The same system may be used to make a master cam for a drum cam, as shown in Fig. 39. Note, however, that the work is set 23 degrees 34 minutes from the vertical position, while the cutter inclines at right angles to, instead of parallel with, the axis of the mandrel. The same combinations of gears would be used if the drum cam action were similar to the one which we have discussed. The exceedingly low cost of making master cams by this method makes it profitable to provide a master cam for cutting the groove in a single cam.

Special Cutter for Finishing Grooved Cams

A source of constant annoyance in milling grooves in cast iron cams lies in the fact that finishing cutters quickly wear and become under size. They must then be laid aside or used for taking the roughing cuts, while a new cutter of full size is used for finishing. We will not discuss the practice of putting a piece of paper in the collet to make the small cutter run out of true. Another source of trouble, even with cutters with spiral flutes, is the tendency of the cutter to chatter, unless it is perfectly ground and all other conditions are exactly right. Still a third trouble is in the tendency of the cutter to cut more on one side than on the other and to dig out stock in spots in the groove.

In Fig. 40 is shown an extremely simple tool, the usefulness of which cannot be overestimated for finishing grooves in cast iron cams. It is a piece of tool steel, suitably machined to mount on an arbor. It is turned on the outside, with enough stock left on for grinding, after which the spiral grooves shown in the developed surface are milled with an angular cutter. The piece is then hardened and ground to size. The cam groove which we are to finish is roughed out from 0.002 inch to 0.012 inch below size; the roughing cutter is removed from the spindle of the cam cutting machine, and this special tool is mounted in its place. The cam is then fed against the tool until the tool reaches the bottom, when the cam is turned one complete revolution. The tool will leave a true groove exactly the right size, and without chatter marks or hollows.

By reason of the form of the cutting or scraping edges, it will outlast many ordinary cutters. Used in connection with it, a single roughing cutter may be repeatedly sharpened before it becomes too small for good results.

CHAPTER V

SUGGESTIONS IN CAM MAKING

In the present chapter are collected a number of suggestions for the laying out and making of cams, together with a discussion on the shape of cam rollers for cylinder cams. These suggestions have been contributed from time to time to the columns of MACHINERY. The names of the persons who originally contributed the matter here selected, have been given in notes at the foot of the pages, together with the month and year when these articles appeared.

Making Master Cams

The method of originating cylindrical master cams, which is described in the following paragraphs, has been used successfully in a shop where considerable of this work is done. A development of the

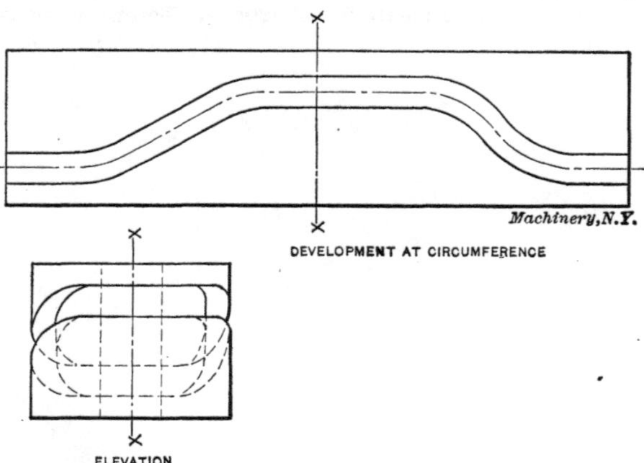

Fig. 41. Master Cam and its Development

cam at the surface of the cylinder is provided by the draftsman. If the cam is smaller than 2½ or 3 inches diameter, or has unusually steep pitches in its make-up, the development should be laid out for a diameter two or three times larger than that of the desired cam.

Suppose it is desired to make a master for the cam shown in Fig. 41. The first step is to make a template to match the development shown in the drawing. This template may be made of mild steel, of a thickness depending upon the diameter to which it is to be bent, as described later. It may be fitted to the drawing with cold chisel and file, or, if considerable accuracy is desired in the throw, the template may be held in the milling machine vise, and the straight surfaces finished to the graduations. This template, shown in Fig. 42, is made for one side of the cam groove only.

The next step is to turn up a piece of steel or cast iron, as shown at *B*, Fig. 43, to such a diameter that when the template *A* is wrapped around it, as shown, the ends will just barely meet. This diameter is about the thickness of the plate less than the diameter to which the development was laid out, but it should be left a little larger and then fitted. The plate is now clasped around the body, with the back edge pushed close up against the shoulder to insure proper alignment

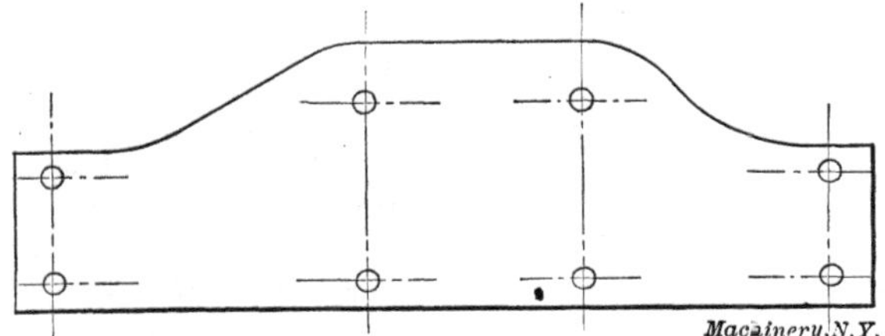

Fig. 42. Template for Making Master Cam in Fig. 41

of the working surface of the cam. If any difficulty is experienced in this wrapping process, a circular strap may be bent up with projecting ends as shown in dotted lines at *C*; with the aid of a clamp *D* the template may be stretched around smoothly. The template and the body may now be drilled and tapped for screws, as shown, and for dowels as well, if found necessary.

Scribe the contour of the cam onto the body *B*, remove the template, place the body on an arbor between the index centers of the milling machine, and take away the stock for about ⅛ inch deep, or so, 1/16

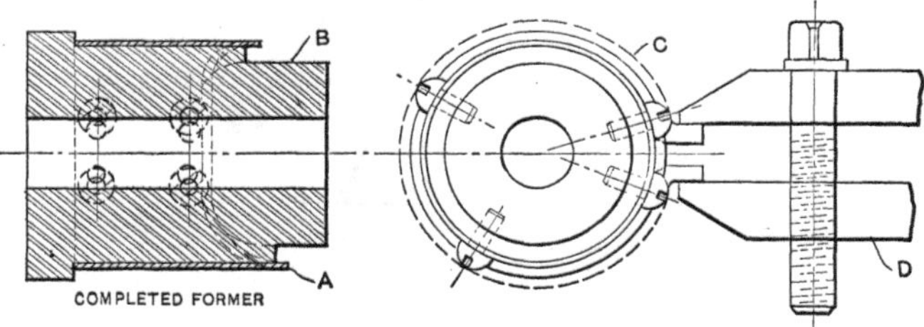

Fig. 43. Template Secured on Mandrel for Making Former

inch back of the scribed line. This, as shown in Fig. 43, is for the purpose of providing a clearance underneath the working edge of the template. The template may now be placed in position on the body once more, and fastened there. The arrangement is now ready for use as a former for making a master cam.

Fig. 44 shows a milling machine arranged for cam cutting. *E* is a casting made to grip the finished face of the column, and carrying an adjustable block *F*. Cam roll *G* is pivoted on a post which is adjustable in and out in block *F*. Our former *H*, and master cam blank *I*, are mounted, as shown, on an arbor between the index centers. By working the index worm crank, and the longitudinal feed together, roll *G* may be made to follow the outline of former *H* in such a manner that the end mill will cut the desired groove in cam blank *I*. A slightly smaller mill may be used for a roughing cut, but it goes without saying that the roll and the finishing mill must be of about the same size if a true copy of the template is desired. It will be found

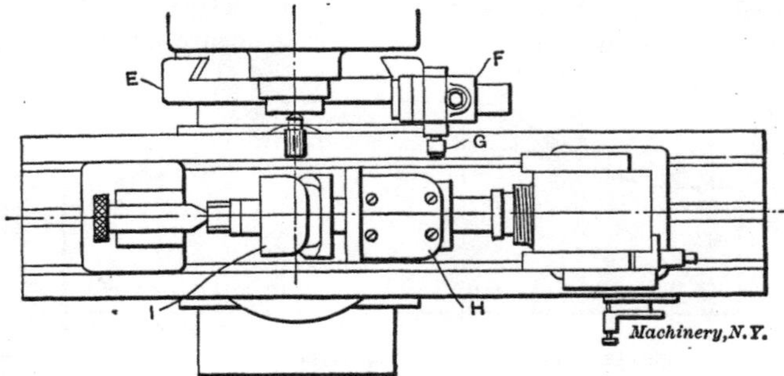

Fig. 44. Arrangement of Milling Machine when Using the Former

easier to follow the outline with the roll if the steeper curves are traced down rather than up.

A fairly good cam cutting machine for making copies of the master cam *I* may be improvised by using the attachment *E*, *F*, *G* in a rack feed machine. It might also be feasible to connect the index worm

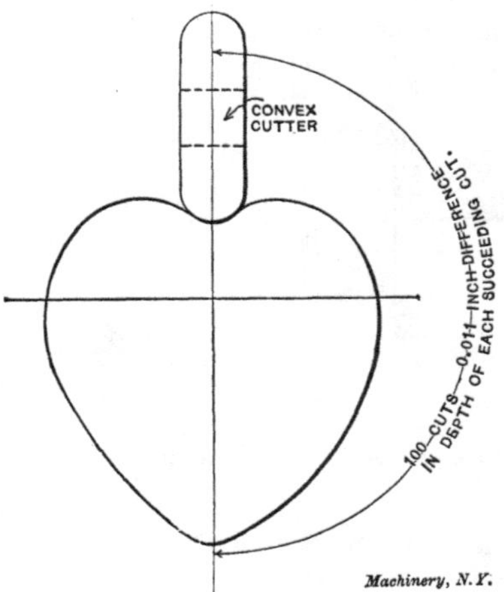

Fig. 45. Method of Cutting Cams

with the telescopic feed shaft so as to give a power feed to the contrivance. To insure accurate cams, the arrangement for holding the tool must be made stiff enough to move the table without much spring, or the table must be weighted, so as to bring the pressure of the roll constantly against one face of the master cam.*

* R. E. Flanders, July, 1904.

Simple Method for Cutting Cams Accurately

Cams are generally laid out with dividers, machined and filed to the line. But for a cam that must advance a certain number of thousandths per revolution of spindle this divider method is not accurate. Cams are easily and accurately made in the following manner. For illustration, let us make the heart cam in Fig. 45. The throw of this cam is 1.1 inch. Now, by setting the index on the miller to cut 200 teeth and also dividing 1.1 inch by 100 we find that we have 0.011 inch to recede from the cam center for each cut across the cam. Placing the cam securely on an arbor, and the latter between the

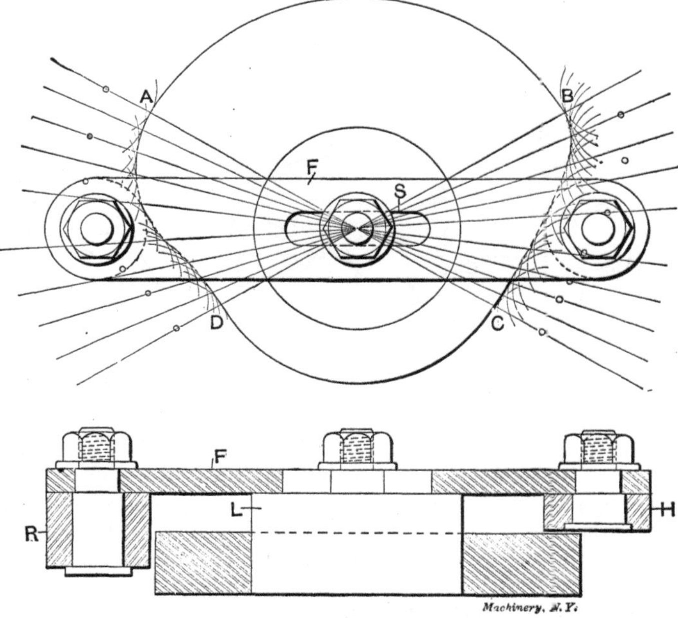

Fig. 46. Device for Correctly Laying Out Cams for Cam-Actuated Press

centers of the milling machine, and using a convex cutter, set the proper distance from the center of the arbor, we make the first cut across the cam. Then, by lowering the milling machine knee 0.011 inch and turning the index pin the proper number of holes on the index plate, we take the next cut and so on. Each cut should be marked on paper so that there will be no mistake as to number of cuts taken; when 100 cuts have been made the knee must be raised in order to complete the opposite side of the cam.*

Device for Laying Out the Cams of a Cam-Actuated Press

The cams which actuate the cutting or drawing slide of a double acting cam-press are different from other cams, inasmuch as each one

* F. E. Shailor, March, 1907.

actuates two rollers which are a certain fixed distance apart from each other. In order to avoid back-lash or springing of the connecting-rods, a fault which is to be found in most cam presses, it is evident that the rollers must both touch the face of the cam at all times. In Fig. 47 is shown the ordinary method of laying out such cams; this cut also shows the fact that this ordinary method does not accomplish the end desired. We see that in this cam both curves which give to the slide its up and down motion are constructed with the same radii, which clearly must give a curve that is faulty at certain points. The one main feature that our cam must possess can be expressed as follows: Two rollers of equal diameters, which are a certain fixed dis-

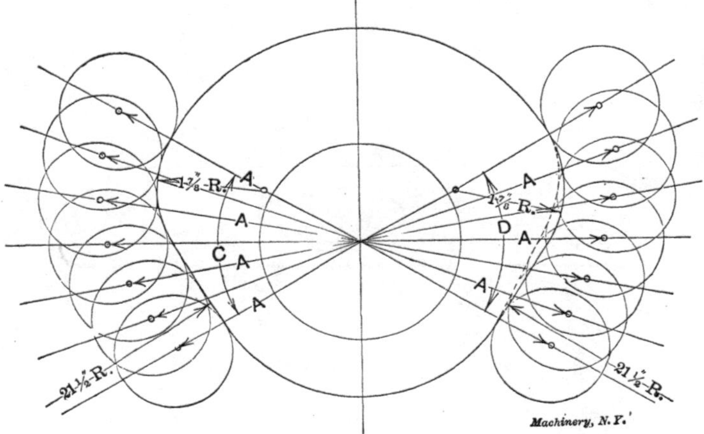

Fig. 47. Ordinary Method of Laying Out Cams

tance (*A* in Fig. 47) apart, on a line passing through center of cam, must always tangent the cam while the cam makes its revolution. Turning to Fig. 47, we see that the curve which spans angle *C* and the dotted curve which spans angle *D* accomplish this object. A little reflection will convince one that this curve cannot be constructed absolutely correct by giving the radii for both the up stroke and down stroke curve, owing to the fact that the shape of one is entirely dependent on the shape of the other.

We can, however, give the radii for one curve, and construct the other curve from it by the aid of the following device. It is assumed that in most cases it will be economical to cut a master cam, and use this for cutting the others. However, where only a few cams are to be cut, it will be well to construct one with the aid of our device, and use this one as a template for the others. Fig. 46 shows the device mentioned. First, cut the two arcs, *A B* and *D C*, which of course are perfect circular arcs of given radii, and also cut the curve *A D* from given radii. Then place center plug *L* into center hole of cam and fasten bar *F* onto *L*. Bar *F* has two rollers, *R* and *H*, fastened in such a way that their center distance is equal to the center distance of the cam rollers in the cam press in which the cams are to be used. The rollers *R* and *H* have the same diameter as the cam rollers in the press. We now keep the roller *R* against the cam along the curve *A D* and follow this curve along its entire length. Center plug *L* will always keep the line connecting *R* and *H* in the center of the cam, and slot *S* enables us to follow the curvature of *A D*. By scratching the outline of roller *H* on the cam blank at very short distances apart, we will have a full outline on the cam blank, which must indicate the absolute curvature of *B C*. This curvature must possess all the qualifications previously set forth as absolutely indispensable for a correct cam-press cam. A cam or set of cams laid out in this manner will silence one of the principal objections usually raised against a cam-actuated press, *viz.*, back-lash or springing of the cam roller connecting-rods.*

Shape of Rolls for Cylinder Cams

The grooves and rolls for cylindrical cams are made in various ways, more or less suitable for the work to be done. Fig. 48 shows a

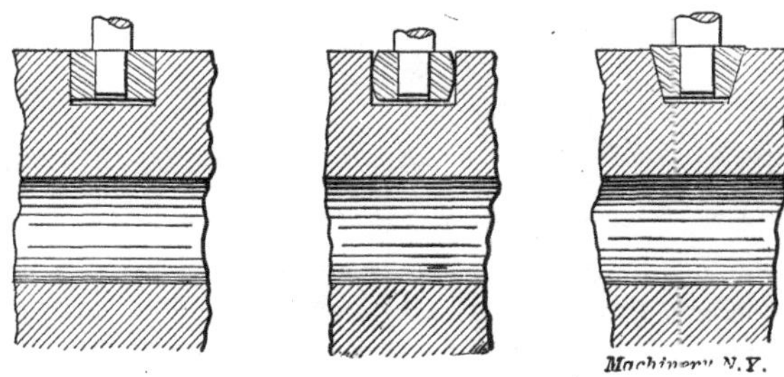

Fig. 48 Fig. 49 Fig. 50
Different Forms of Cam Rolls

straight roll and groove, Fig. 49, a roll with a rounded surface in a straight sided groove, and Fig. 50 a beveled roll and groove. In Fig. 48 the action of the roll is faulty, because of the varying surface speed of the cam at the top and bottom of the groove, due to its varying radial distance from the center line. This causes excessive wear and friction, especially in a quick running cam with steep pitches. For such cases, if the duty is light, the arrangement shown in Fig. 49 is better, as the roll has but a very small bearing surface, and is thus unaffected by a varying radial distance. For heavy work, however, the small bearing area is quickly worn down, and the roll presses a groove into the side of the cam as well, destroying the accuracy of the movement, and allowing a great amount of back-lash.

In Fig. 50 the conical shape is given to the roll with the idea of giving it a true rolling action in the groove. In most cases where this shape is used, however, the lines of the sides of the roll appear to converge, to the center line of the cam, as shown in the figure. If the groove were a plain circumferential one, it would give a perfect action, like that of the pitch cones of two bevel gears rolling on each

* E. E. Eisenwinter, July, 1907.

other. If the motion were in a line with the axis of the cam, without any circular movement, conditions would be perfect in Fig. 48. It is evident that in intermediate conditions, the groove must be given a shape intermediate between the two. In many cams of this variety the heavy duty comes on a section of the cam which is of nearly even pitch and of considerable length. In such a case it is best to proportion the shape of the roll to work correctly during the important part of the cycle, letting it go as it will at other times.

In Fig. 51, *b* is the circumferential distance on the surface of the cam, which includes the movement we desire to fit the roll to. The

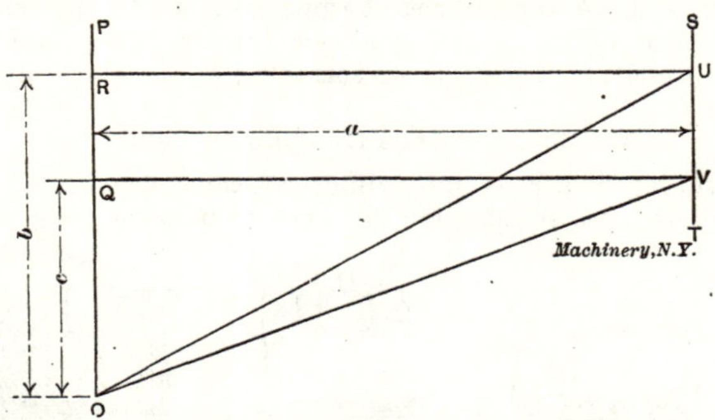

Fig. 51. Diagram Showing Method of Finding Shape of Cam Rolls

throw of the cam for this circumferential movement is *a*. Line *OU* will then be a development of the movement of the cam roll during the given part of the cycle, and *c* is the movement corresponding to *b*, but on a circle whose diameter is that of the cam at the bottom of the groove. With the same throw *a* as before, the line *OV* will be a development of the cam at the bottom of the groove. *OU* then is the length of the helix traveled by the top of the roll, while *OV* is the amount of travel at the bottom of the groove. If then the top width and the bottom width of the groove be made proportional to *OU* and *OV*, the shape will be suitable to give the result we are seeking.*

* R. E. Flanders, December, 1904.

Instructions and Tables
for
Cutting Screw Machine Cams
on the
Milling Machine

Extract from our "Practical Treatise on Milling and Milling Machines"

BROWN & SHARPE MFG. CO.
PROVIDENCE, R. I.
U. S. A.

Form 341 DR

25C 4-19

A METHOD OF MILLING SCREW MACHINE CAMS

Using a Spiral Head and Vertical Spindle Milling Attachment

Peripheral cams for Brown & Sharpe Automatic Screw Machines can be cut upon a milling machine, and a far more satisfactory job can be obtained than is possible by drilling around the outline on a cam blank, breaking it off and then milling or filing to a line.

Either a Plain or Universal Milling Machine can be employed, which must be equipped with a spiral head and a vertical spindle milling attachment. The spiral head is geared to the table feed screw, the same as in cutting ordinary spirals, and the cam blank is fastened to the end of the index spindle. An end mill is used in the vertical spindle milling attachment, which is set in each case to mill the periphery of the cam at right angles to its sides, or, in other words, the axes of the spiral head spindle and attachment spindle must always be parallel to mill cams according to this method. The cutting is done by the teeth on the periphery of the end mill. The principle of this method is as follows: Suppose the spiral head is elevated to 90°, or at exact right angles to the surface of the table (See Fig. 1), and is geared for any given lead. It is then apparent that, as the table advances and the blank is turned, the distance between the axes of the index spindle and attachment spindle becomes less. In other words, the cut becomes deeper and the radius of the cam is shortened, producing a spiral lobe, the lead of which is the same as that for which the machine is geared.

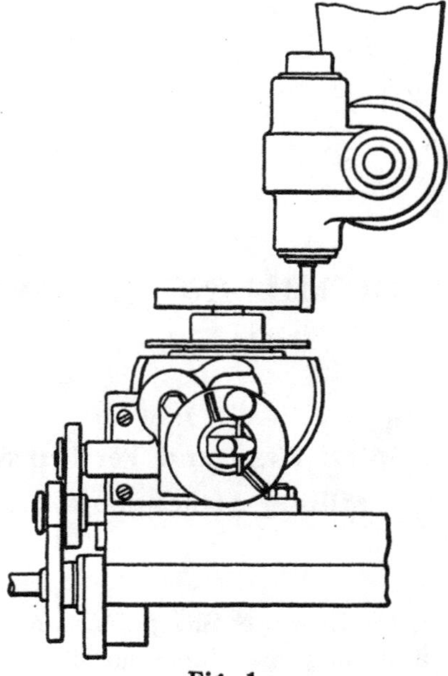

Fig. 1

Fig. 2

Now, suppose the same gearing is retained and the spiral head is set at zero, or parallel to the surface of the table (see Fig. 2). It is apparent, also, that the axes of the index spindle and attachment spindle are parallel to one another. Therefore, as the table advances, and the blank is turned, the distance between the axes of the index spindle and attachment spindle remains the same. As a result, the periphery of the blank, if milled, is concentric or the lead is 0.

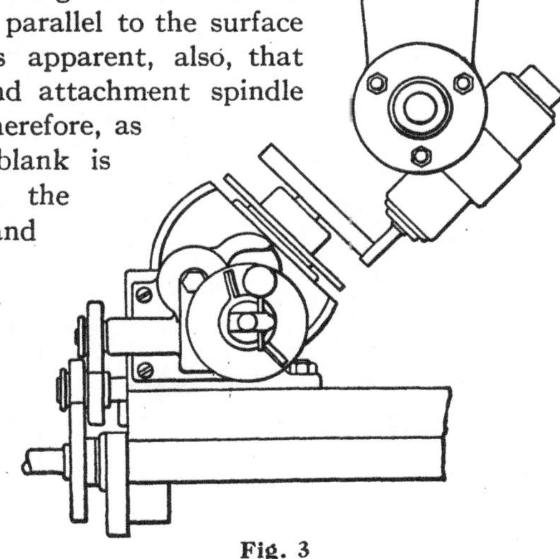

Fig. 3

If, then, the spiral head is elevated to any angle between zero and 90° (see Fig. 3), the amount of lead given to the cam will be between that for which the machine is geared and 0. Hence it is clear that a very large range of different leads can be obtained with one set of change gears, and the problem of milling the lobes of a cam is reduced to a question of finding the angle at which to set the head to obtain any given lead.

In order to illustrate the method of obtaining the correct angle, drawings of two cams to be milled, and data connected with same, are given in Figs. 4 and 5.

It is first necessary to know the lead of the lobes of a cam, that is, the amount of rise of each lobe if continued the full circumference of the cam. This can be obtained from the drawings as follows: For cams where the face is divided into hundredths, as those shown: multiply 100 by the rise of the lobe in inches and divide by the number of hundredths of circumference occupied by the lobe. For cams that are figured in degrees of circumference: multiply 360 by the rise of the lobe in inches and divide by the number of degrees of circumference occupied by the lobe. Taking Fig. 4 for example, we have a cam of one lobe which extends through 91 hundredths of the circumference, and has a rise .178″. Then $\frac{100 \times .178''}{91} = .1956$ lead of lobe, or .196″, which is near enough for all practical purposes.

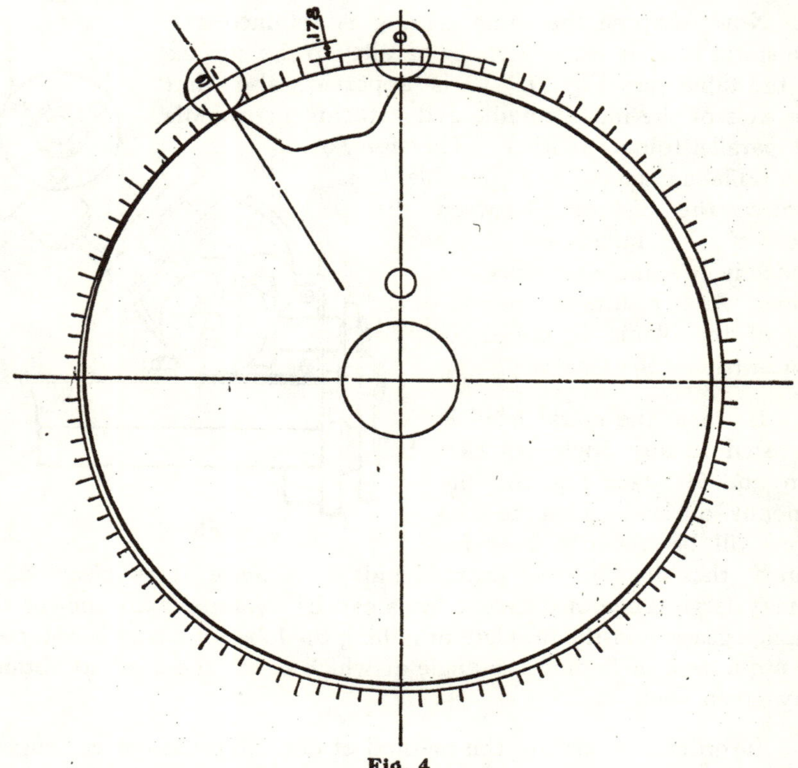

Fig. 4

As a .196″ lead is much less than .67″, which is the shortest lead* regularly obtainable on the milling machine, the change gears that will give a lead of .67″ may be used, and then the angle of the head can be adjusted so that a lead of .196″ will be obtained on the cam lobe with these change gears. The rule for this is:

Divide the given lead of the cam lobe by a lead obtainable on the machine, and the result is the sine of the angle at which to set the head.

Continuing the calculation for the lobe of the cam in Fig. 4, we therefore have:

$$\frac{.196''}{.67} = .29253$$

Hence, .29253 is the sine of the correct angle. Referring to a table of sines and cosines we find that .29253 is very near .29265, which

*By the use of our short lead spiral attachment much shorter leads than .67″ are obtainable.

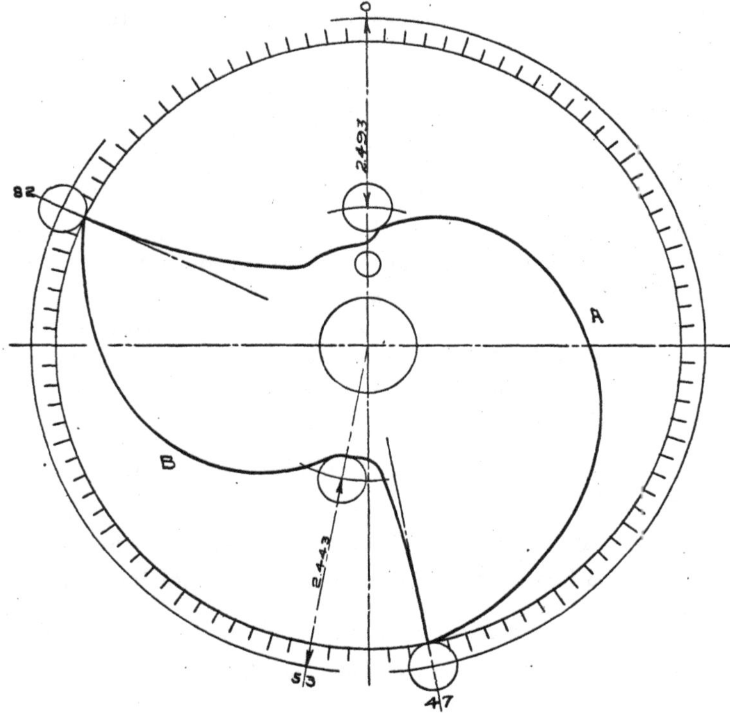

Fig. 5

is the sine of an angle of 17° and 1′. As the spiral head is not graduated closer than quarter degrees, it will be satisfactory to elevate the head just a hair over 17°; then, with the gearing for a lead of .67″, a lead of .196″ will be obtained.

The minute errors between the actual lead .1956″ and .196″, and in the sines and angles of this calculation can be safely ignored, as it is not possible in practice to work very much closer than we have outlined.

The portion of the periphery of the cam from 91 hundredths to zero, represents a clearance of the cutting tool prior to the beginning of the throw. It is usually milled to a line, or drilled, broken out, and filed.

In Fig. 5, we have a cam with two lobes, one, A, having a rise of 2.493″ in 47 hundredths, and the other, B, having a rise of 2.443″ in 29 hundredths. On cams such as this, where it is necessary to remove considerable stock, it is usually the practice to first outline the approximate shape of the lobes on the blank and drill and break off the surplus stock.

Following the same method of figuring to find the lead of the lobes on this cam, we have: $\frac{100 \times 2.493''}{47} = 5.304''$ lead for lobe A, and $\frac{100 \times 2.443''}{29} = 8.424''$ lead for lobe B.

Where there are two or more lobes on a cam, the machine is geared for a lead slightly longer than the longest one required, which in this case is 8.424″, then the other lobes are milled without changing the gears. Referring to the Table of Leads furnished with the milling machine, we find a lead of 8.437″, which is slightly larger than 8.424″. This gearing is, therefore, accepted, and it is required to find the sine of the angle at which to set the head for lobe B.

$\frac{8.424}{8.437} = .99846$ sine of angle at which to set head. Looking at a table of sines and cosines, .99846 is found to be the sine of an angle of 86° and 49′. The head is, therefore, set at a trifle over $86\frac{3}{4}$°.

When lobe B has been milled, the head is set for lobe A.

$\frac{5.304}{8.437} = .62865$ sine of an angle at which to set head. Referring again to the table of sines and cosines, we find that .62865 is very near to .62864, which is the sine of an angle of 38° and 57′. The head is, therefore, set slightly under 39° for this lobe.

The other portions of the periphery of this cam are formed up either by filing to a line before the blank is put on the milling machine, or by milling to the line after the lobes have been formed.

Whenever possible, the job should be set up so that the end mill will cut on the lower side of the blank, as this brings the mill and table nearer together and makes the job more rigid. It also prevents chips from accumulating, and enables the operator to better see any lines that may be laid out on the face of the cam.

When the lead is over two inches the automatic feed can be used, but when the lead is less than 2 inches the job should be fed by hand, with the index crank.

By the use of the calculations just given, we have compiled the tables on following pages that give a wide range of leads from 0 to 20″ that can be obtained with the spiral head in the manner described. These tables will be found useful, as they give all data and settings without the necessity of figuring.

TABLES OF LEADS FOR CAM LOBES

Obtained with Spiral Head and a Vertical Spindle Milling Attachment Set at an Angle

In compiling these tables, we have employed only the change gears furnished with our spiral head, and the combinations of change gears in each case will reach without interfering. The practical leads obtainable with each set of change gears have been grouped together so that when a machine is set for any lead, and it is desired to change to another lead, the operator can quickly determine whether the required lead is available without changing the gears already on. As this is often the case in this work, the saving in time that is effected is readily appreciated.

A selection of leads from 0 to 20″ is listed, and it should be understood that these are the leads or amount of rise in a complete circle, not the amount of rise of a lobe in a fractional part of the circumference. From the amount of rise of the lobe it will be necessary before using these tables to calculate the lead or rise if the lobe were continued the full circumference. This is easily found as explained on page 5.

In using these tables to set up a machine to mill any required lead, the column under the heading "Approximate Lead" is first followed down until the range of leads is found which embraces the required one. Then follow the horizontal line across until the nearest dimension to the exact lead required is found. At the top of the column containing this dimension will be found the required combination of change gears, and in the next two columns at the right, and in line with the dimension selected, will be found the angles at which to set the spiral head and vertical milling attachment.

Example: Required, the change gears and angles at which to set the spiral head and vertical milling attachment for a lead of .1476″.

Following down the first column we find .145-50, which embraces the required lead. Following this line across horizontally we find .1474″, which is sufficiently near to .1476″ for all practical purposes. At the top of the column containing .1474″ is the proper combination of change gears, 24, 86, 32, and 100, and in the two columns at the right and in line with .1474″ are the necessary angles; 9½° for spiral head, and 80½° for vertical milling attachment.

When the machine is already set for a given lead and it is desired to know whether another required lead can be obtained without changing the gears, proceed as follows:

Example: Machine is set with a combination of gears, 24, 72, 32, and 86, and a lead of .1080″ is required.

Follow down the column of exact leads that are given under the combination of change gears for which the machine is set until .1081″ is found. This is sufficiently near to .1080″ for all practical purposes. Hence it is possible to obtain this lead without changing the gears, by setting the spiral head at 5° and the vertical milling attachment at 85°.

In milling cams in this way an angle of greater than 80° with the spiral head, which is the greatest angle listed in these tables, should be avoided to prevent going beyond the range of the spiral head.

A vertical spindle milling attachment with offset spindle, is preferable for this work, as it will reach nearer to the spiral head spindle when milling small cams with the head set nearly vertical.

We also manufacture an extension by the use of which the spiral head can be moved farther in on the table to bring the spiral head and vertical spindle attachment spindles nearer together. This extension is furnished on special order.

The standard end mill is of sufficient length for practically all leads on ordinary screw machine cams, for long leads usually extend over only a partial turn of the cam.

The mill should be of the same diameter as the roll to be used with the cam.

Approximate Lead	Gear on Worm 24 / 1st on Stud 86 / 2nd on Stud 24 / Gear on Screw 100	Angle to Set Spiral Head (Degrees)	Angle to Set Vertical Att. (Degrees)	Gear on Worm 24 / 1st on Stud 86 / 2nd on Stud 28 / Gear on Screw 100	Angle to Set Spiral Head (Degrees)	Angle to Set Vertical Att. (Degrees)	Gear on Worm 24 / 1st on Stud 86 / 2nd on Stud 32 / Gear on Screw 100	Angle to Set Spiral Head (Degrees)	Angle to Set Vertical Att. (Degrees)	Gear on Worm 24 / 1st on Stud 72 / 2nd on Stud 28 / Gear on Screw 86	Angle to Set Spiral Head (Degrees)	Angle to Set Vertical Att. (Degrees)
.000-05	.0029	¼	89¾	.0034	¼	89¾	.0039	¼	89¾	.0047	¼	89¾
.005-10	.0088	¾	89¼	.0068	½	89½	.0078	½	89½	.0095	½	89½
.010-15	.0117	1	89	.0102	¾	89¼	.0117	¾	89¼	.0142	¾	89¼
.015-20	.0175	1½	88½	.0170	1¼	88¾	.0195	1¼	88¾	.0189	1	89
.020-25	.0234	2	88	.0205	1½	88½	.0234	1½	88½	.0237	1¼	88¾
.025-30	.0263	2¼	87¾	.0273	2	88	.0273	1¾	88¼	.0284	1½	88½
.030-35	.0321	2¾	87¼	.0341	2½	87½	.0312	2	88	.0331	1¾	88¼
.035-40	.0351	3	87	.0375	2¾	87¼	.0390	2½	87½	.0379	2	88
.040-45	.0409	3½	86½	.0443	3¼	86¾	.0428	2¾	87¼	.0426	2¼	87¾
.045-50	.0497	4¼	85¾	.0477	3½	86½	.0467	3	87	.0473	2½	87½
.050-55	.0526	4½	85½	.0511	3¾	86¼	.0545	3½	86½	.0521	2¾	87¼
.055-60	.0584	5	85	.0579	4¼	85¾	.0584	3¾	86¼	.0568	3	87
.060-65	.0642	5½	84½	.0613	4½	85½	.0623	4	86	.0615	3¼	86¾
.065-70	.0671	5¾	84¼	.0681	5	85	.0662	4¼	85¾	.0662	3½	86½
.070-75	.0700	6	84	.0749	5½	84½	.0739	4¾	85¼	.0710	3¾	86¼
.075-80	.0758	6½	83½	.0782	5¾	84¼	.0778	5	85	.0757	4	86
.080-85	.0846	7¼	82¾	.0816	6	84	.0817	5¼	84¾	.0804	4¼	85¾
.085-90	.0875	7½	82½	.0884	6½	83½	.0895	5¾	84¼	.0899	4¾	85¼
.090-95	.0932	8	82	.0918	6¾	83¼	.0933	6	84	.0946	5	85
.095-00	.0961	8¼	81¾	.0952	7	83	.0972	6¼	83¾	.0993	5¼	84¾
.100-05	.1048	9	81	.1020	7½	82½	.1011	6½	83½	.1040	5½	84½
.105-10	.1077	9¼	80¾	.1053	7¾	82¼	.1088	7	83	.1087	5¾	84¼
.110-15	.1106	9½	80½	.1121	8¼	81¾	.1127	7¼	82¾	.1134	6	84
.115-20	.1163	10	80	.1155	8½	81½	.1166	7½	82½	.1181	6¼	83¾
.120-25	.1221	10½	79½	.1222	9	81	.1204	7¾	82¼	.1228	6½	83½
.125-30	.1250	10¾	79¼	.1289	9½	80½	.1281	8¼	81¾	.1275	6¾	83¼
.130-35	.1336	11½	78½	.1323	9¾	80¼	.1320	8½	81½	.1322	7	83
.135-40	.1393	12	78	.1356	10	80	.1397	9	81	.1369	7¼	82¾
.140-45	.1421	12¼	77¾	.1424	10½	79½	.1435	9¼	80¾	.1416	7½	82½
.145-50	.1450	12½	77½	.1491	11	79	.1474	9½	80½	.1463	7¾	82¼

Approximate Lead	Gear on Worm 24 / 1st on Stud 72 / 2nd on Stud 32 / Gear on Screw 86	Angle to Set Spiral Head (Degrees)	Angle to Set Vertical Att. (Degrees)	Gear on Worm 24 / 1st on Stud 72 / 2nd on Stud 40 / Gear on Screw 100	Angle to Set Spiral Head (Degrees)	Angle to Set Vertical Att. (Degrees)	Gear on Worm 24 / 1st on Stud 48 / 2nd on Stud 28 / Gear on Screw 100	Angle to Set Spiral Head (Degrees)	Angle to Set Vertical Att. (Degrees)	Gear on Worm 24 / 1st on Stud 64 / 2nd on Stud 40 / Gear on Screw 100	Angle to Set Spiral Head (Degrees)	Angle to Set Vertical Att. (Degrees)
.000-05												
.005-10	.0054	¼	89¾	.0058	¼	89¾	.0061	¼	89¾	.0065	¼	89¾
.010-15	.0108	½	89½	.0116	½	89½	.0122	½	89½	.0131	½	89½
.015-20	.0162	¾	89¼	.0174	¾	89¼	.0183	¾	89¼	.0196	¾	89¼
.020-25	.0216	1	89	.0233	1	89	.0244	1	89			
.025-30	.0271	1¼	88¾	.0291	1¼	88¾				.0262	1	89
.030-35	.0327	1½	88½	.0349	1½	88½	.0305	1¼	88¾	.0327	1¼	88¾
.035-40	.0379	1¾	88¼				.0367	1½	88½	.0393	1½	88½
.040-45	.0433	2	88	.0407	1¾	88¼	.0428	1¾	88¼			
.045-50	.0487	2¼	87¾	.0465	2	88	.0489	2	88	.0458	1¾	88¼
.050-55	.0541	2½	87½	.0524	2¼	87¾				.0524	2	88
.055-60	.0595	2¾	87¼	.0582	2½	87½	.0550	2¼	87¾	.0589	2¼	87¾
.060-65	.0649	3	87	.0640	2¾	87¼	.0611	2½	87½			
.065-70				.0698	3	87	.0672	2¾	87¼	.0654	2½	87½
.070-75	.0703	3¼	86¾				.0733	3	87	.0720	2¾	87¼
.075-80	.0757	3½	86½	.0756	3¼	86¾	.0794	3¼	86¾	.0785	3	87
.080-85	.0811	3¾	86¼	.0814	3½	86½						
.085-90	.0865	4	86	.0872	3¾	86¼	.0855	3½	86½	.0850	3¼	86¾
.090-95	.0919	4¼	85¾	.0930	4	86	.0916	3¾	86¼	.0916	3½	86½
.095-00	.0973	4½	85½	.0988	4¼	85¾	.0978	4	86	.0981	3¾	86¼
.100-05	.1027	4¾	85¼	.1046	4½	85½	.1038	4¼	85¾	.1046	4	86
.105-10	.1081	5	85				.1098	4½	85½			
.110-15	.1135	5¼	84¾	.1104	4¾	85¼				.1112	4¼	85¾
.115-20	.1188	5½	84½	.1162	5	85	.1159	4¾	85¼	.1177	4½	85½
.120-25	.1242	5¾	84¼	.1220	5¼	84¾	.1220	5	85	.1242	4¾	85¼
.125-30	.1296	6	84	.1278	5½	84½	.1281	5¼	84¾			
.130-35				.1336	5¾	84¼	.1342	5½	84½	.1307	5	85
.135-40	.1350	6¼	83¾	.1393	6	84				.1372	5¼	84¾
.140-45	.1404	6½	83½				.1403	5¾	84¼	.1438	5½	84½
.145-50	.1458	6¾	83¼	.1452	6¼	83¾	.1463	6	84			

LEADS FROM .150 TO .300

Approximate Lead	Gear on Worm 24, 1st on Stud 86, 2nd on Stud 24, Gear on Screw 100	Angle to Set Spiral Head (Degrees)	Angle to Set Vertical Att. (Degrees)	Gear on Worm 24, 1st on Stud 86, 2nd on Stud 28, Gear on Screw 100	Angle to Set Spiral Head (Degrees)	Angle to Set Vertical Att. (Degrees)
.150-55	.1535	13¼	76¾	.1524	11¼	78¾
.155-60	.1564	13½	76½	.1591	11¾	78¼
.160-65	.1621	14	76	.1624	12	78
.165-70	.1677	14½	75½	.1691	12½	77½
.170-75	.1734	15	75	.1724	12¾	77¼
.175-80	.1790	15½	74½	.1757	13	77
.180-85	.1847	16	74	.1824	13½	76½
.185-90	.1875	16¼	73¾	.1890	14	76
.190-95	.1903	16½	73½	.1923	14¼	75¾
.195-00	.1959	17	73	.1989	14¾	75¼
.200-05	.2014	17½	72½	.2022	15	75
.205-10	.2070	18	72	.2055	15¼	74¾
.210-15	.2126	18½	71½	.2121	15¾	74¼
.215-20	.2181	19	71	.2154	16	74
.220-25	.2236	19½	70½	.2219	16½	73½
.225-30	.2291	20	70	.2284	17	73
.230-35	.2346	20½	69½	.2349	17½	72½
.235-40	.2373	20¾	69¼	.2382	17¾	72¼
.240-45	.2401	21	69	.2415	18	72
.245-50	.2482	21¾	68¼	.2479	18½	71½
.250-55	.2510	22	68	.2544	19	71
.255-60	.2564	22½	67½	.2576	19¼	70¾
.260-65	.2618	23	67	.2640	19¾	70¼
.265-70	.2698	23¾	66¼	.2672	20	70
.270-75	.2725	24	66	.2704	20¼	69¾
.275-80	.2778	24½	65½	.2768	20¾	69¼
.280-85	.2831	25	65	.2800	21	69
.285-90	.2884	25½	64½	.2864	21½	68½
.290-95	.2937	26	64	.2927	22	68
.295-00	.2963	26¼	63¾	.2990	22½	67½

Approximate Lead	Gear on Worm 24, 1st on Stud 86, 2nd on Stud 32, Gear on Screw 100	Angle to Set Spiral Head (Degrees)	Angle to Set Vertical Att. (Degrees)	Gear on Worm 24, 1st on Stud 72, 2nd on Stud 28, Gear on Screw 86	Angle to Set Spiral Head (Degrees)	Angle to Set Vertical Att. (Degrees)
.150-55	.1512	9¾	80¼	.1510	8	82
.155-60	.1551	10	80	.1557	8¼	81¾
.160-65	.1627	10½	79½	.1604	8½	81½
.165-70	.1666	10¾	79¼	.1650	8¾	81¼
.170-75	.1704	11	79	.1744	9¼	80¾
.175-80	.1780	11½	78½	.1791	9½	80½
.180-85	.1818	11¾	78¼	.1837	9¾	80¼
.185-90	.1857	12	78	.1884	10	80
.190-95	.1933	12½	77½	.1931	10¼	79¾
.195-00	.1971	12¾	77¼	.1977	10½	79½
.200-05	.2009	13	77	.2024	10¾	79¼
.205-10	.2085	13½	76½	.2070	11	79
.210-15	.2122	13¾	76¼	.2117	11¼	78¾
.215-20	.2160	14	76	.2163	11½	78½
.220-25	.2231	14½	75½	.2210	11¾	78¼
.225-30	.2274	14¾	75¼	.2256	12	78
.230-35	.2311	15	75	.2302	12¼	77¾
.235-40	.2386	15½	74½	.2395	12¾	77¼
.240-45	.2424	15¾	74¼	.2441	13	77
.245-50	.2462	16	74	.2487	13¼	76¾
.250-55	.2536	16½	73½	.2533	13½	76½
.255-60	.2574	16¾	73¼	.2579	13¾	76¼
.260-65	.2648	17¼	72¾	.2625	14	76
.265-70	.2685	17½	72½	.2671	14¼	75¾
.270-75	.2722	17¾	72¼	.2717	14½	75½
.275-80	.2760	18	72	.2762	14¾	75¼
.280-85	.2833	18½	71½	.2808	15	75
.285-90	.2871	18¾	71¼	.2854	15¼	74¾
.290-95	.2907	19	71	.2900	15½	74½
.295-00	.2981	19½	70½	.2991	16	74

Approximate Lead	Gear on Worm 24, 1st on Stud 72, 2nd on Stud 32, Gear on Screw 86	Angle to Set Spiral Head (Degrees)	Angle to Set Vertical Att. (Degrees)	Gear on Worm 24, 1st on Stud 72, 2nd on Stud 40, Gear on Screw 100	Angle to Set Spiral Head (Degrees)	Angle to Set Vertical Att. (Degrees)
.150-55	.1511	7	83	.1509	6½	83½
.155-60	.1565	7¼	82¾	.1568	6¾	83¼
.160-65	.1619	7½	82½	.1625	7	83
.165-70	.1672	7¾	82¼	.1683	7¼	82¾
.170-75	.1726	8	82	.1741	7½	82½
.175-80	.1779	8¼	81¾	.1798	7¾	82¼
.180-85	.1833	8½	81½			
.185-90	.1887	8¾	81¼	.1856	8	82
.190-95	.1942	9	81	.1914	8¼	81¾
.195-00	.1993	9¼	81¾	.1971	8½	81½
.200-05	.2047	9½	81½	.2029	8¾	81¼
.205-10				.2086	9	81
.210-15	.2100	9¾	81¼	.2143	9¼	80¾
.215-20	.2153	10	80			
.220-25	.2207	10¼	79¾	.2201	9½	80½
.225-30	.2260	10½	79½	.2258	9¾	80¼
.230-35	.2313	10¾	79¼	.2316	10	80
.235-40	.2366	11	79	.2373	10¼	79¾
.240-45	.2420	11¼	78¾	.2430	10½	79½
.245-50	.2473	11½	78½	.2487	10¾	79¼
.250-55	.2526	11¾	78¼	.2545	11	79
.255-60	.2579	12	78			
.260-65	.2632	12¼	77¾	.2601	11¼	78¾
.265-70	.2685	12½	77½	.2659	11½	78½
.270-75	.2738	12¾	77¼	.2716	11¾	78¼
.275-80	.2790	13	77	.2773	12	78
.280-85	.2843	13¼	76¾	.2829	12¼	77¾
.285-90	.2896	13½	76½	.2886	12½	77½
.290-95	.2948	13¾	76¼	.2943	12¾	77¼
.295-00				.2999	13	77

Approximate Lead	Gear on Worm 24, 1st on Stud 48, 2nd on Stud 28, Gear on Screw 100	Angle to Set Spiral Head (Degrees)	Angle to Set Vertical Att. (Degrees)	Gear on Worm 24, 1st on Stud 64, 2nd on Stud 40, Gear on Screw 100	Angle to Set Spiral Head (Degrees)	Angle to Set Vertical Att. (Degrees)
.150-55	.1524	6¼	83¾	.1503	5¾	84¼
.155-60	.1585	6½	83½	.1568	6	84
.160-65	.1645	6¾	83¼	.1633	6¼	83¾
.165-70				.1698	6½	83½
.170-75	.1706	7	83			
.175-80	.1767	7¼	82¾	.1763	6¾	83¼
.180-85	.1828	7½	82½	.1828	7	83
.185-90	.1888	7¾	82¼	.1893	7¼	82¾
.190-95	.1948	8	82			
.195-00				.1958	7½	82½
.200-05	.2009	8¼	81¾	.2023	7¾	82¼
.205-10	.2068	8½	81½	.2088	8	82
.210-15	.2130	8¾	81¼			
.215-20	.2190	9	81	.2152	8¼	81¾
.220-25				.2217	8½	81½
.225-30	.2250	9¼	80¾	.2282	8¾	81¼
.230-35	.2311	9½	80½	.2347	9	81
.235-40	.2371	9¾	79¼			
.240-45	.2431	10	80	.2411	9¼	80¾
.245-50	.2491	10¼	79¾	.2476	9½	80½
.250-55				.2540	9¾	80¼
.255-60	.2552	10½	79½			
.260-65	.2612	10¾	79¼	.2605	10	80
.265-70	.2672	11	79	.2669	10¼	79¾
.270-75	.2732	11¼	78¾	.2734	10½	79½
.275-80	.2791	11½	78½	.2798	10¾	79¼
.280-85						
.285-90	.2851	11¾	78¼	.2862	11	79
.290-95	.2911	12	78	.2926	11¼	78¾
.295-00	.2971	12¼	77¾	.2991	11½	78½

Approximate Lead	Gear on Worm 24, 1st on Stud 86, 2nd on Stud 24, Gear on Screw 100	Angle to Set Spiral Head	Angle to Set Vertical Att.	Gear on Worm 24, 1st on Stud 86, 2nd on Stud 28, Gear on Screw 100	Angle to Set Spiral Head	Angle to Set Vertical Att.	Gear on Worm 24, 1st on Stud 86, 2nd on Stud 32, Gear on Screw 100	Angle to Set Spiral Head	Angle to Set Vertical Att.	Gear on Worm 24, 1st on Stud 72, 2nd on Stud 28, Gear on Screw 86	Angle to Set Spiral Head	Angle to Set Vertical Att.
		Degrees	Degrees		Degrees	Degrees		Degrees	Degrees		Degrees	Degrees
.300-05	.3042	27	63	.3021	22¾	67¼	.3018	19¾	70¼	.3036	16¼	73¾
.305-10	.3068	27¼	62¾	.3053	23	67	.3054	20	70	.3082	16½	73½
.310-15	.3145	28	62	.3115	23½	66½	.3127	20½	69½	.3127	16¾	73¼
.315-20	.3197	28½	61½	.3178	24	66	.3164	20¾	69¼	.3173	17	73
.320-25	.3248	29	61	.3240	24½	65½	.3200	21	69	.3218	17¼	72¾
.325-30	.3299	29½	60½	.3271	24¾	65¼	.3273	21½	68½	.3263	17½	72½
.330-35	.3324	29¾	60¼	.3302	25	65	.3345	22	68	.3308	17¾	72¼
.335-40	.3350	30	60	.3364	25½	64½				.3353	18	72
.340-45	.3400	30½	59½	.3425	26	64	.3417	22½	67½	.3443	18½	71½
.345-50	.3451	31	59	.3487	26½	63½	.3489	23	67	.3488	18¾	71¼
.350-55	.3500	31½	58½	.3547	27	63	.3525	23¼	66¾	.3532	19	71
.355-60	.3550	32	58	.3578	27¼	62¾	.3561	23½	66½	.3577	19¼	70¾
.360-65	.3600	32½	57½	.3608	27½	62½	.3632	24	66	.3622	19½	70½
.365-70	.3650	33	57	.3668	28	62	.3667	24¼	65¾	.3666	19¾	70¼
.370-75	.3746	34	56	.3728	28½	61½	.3703	24½	65½	.3711	20	70
.375-80	.3794	34½	55½	.3788	29	61	.3774	25	65	.3755	20¼	69¾
.380-85	.3842	35	55	.3848	29½	60½	.3809	25¼	64¾	.3800	20½	69½
.385-90	.3866	35¼	54¾	.3877	29¾	60¼	.3880	25¾	64¼	.3888	21	69
.390-95	.3937	36	54	.3907	30	60	.3915	26	64	.3932	21¼	68¾
.395-00	.3985	36½	53½	.3967	30½	59½	.3985	26½	63½	.3977	21½	68½
.400-05	.4032	37	53	.4024	31	59	.4020	26¾	63¼	.4021	21¾	68¼
.405-10	.4078	37½	52½	.4083	31½	58½	.4054	27	63	.4065	22	68
.410-15	.4101	37¾	52¼	.4140	32	58	.4123	27½	62½	.4108	22¼	67¾
.415-20	.4170	38½	51½	.4198	32½	57½	.4192	28	62	.4152	22½	67½
.420-25	.4216	39	51	.4227	32¾	57¼	.4227	28¼	61¾	.4239	23	67
.425-30	.4261	39½	50½	.4256	33	57	.4261	28½	61½	.4283	23¼	66¾
.430-35	.4306	40	50	.4313	33½	56½	.4329	29	61	.4326	23½	66½
.435-40	.4351	40½	49½	.4369	34	56	.4397	29½	60½	.4370	23¾	66¼
.440-45	.4439	41½	48½	.4426	34½	55½	.4431	29¾	60¼	.4413	24	66
.445-50	.4482	42	48	.4482	35	55	.4465	30	60	.4456	24¼	65¾

Approximate Lead	Gear on Worm 24, 1st on Stud 72, 2nd on Stud 32, Gear on Screw 86	Angle to Set Spiral Head	Angle to Set Vertical Att.	Gear on Worm 24, 1st on Stud 72, 2nd on Stud 40, Gear on Screw 100	Angle to Set Spiral Head	Angle to Set Vertical Att.	Gear on Worm 24, 1st on Stud 48, 2nd on Stud 28, Gear on Screw 100	Angle to Set Spiral Head	Angle to Set Vertical Att.	Gear on Worm 24, 1st on Stud 64, 2nd on Stud 40, Gear on Screw 100	Angle to Set Spiral Head	Angle to Set Vertical Att.
		Degrees	Degrees		Degrees	Degrees		Degrees	Degrees		Degrees	Degrees
.300-05	.3000	14	76				.3030	12½	77½			
.305-10	.3053	14¼	75¾	.3056	13¼	76¾	.3090	12¾	77¼	.3055	11¾	78¼
.310-15	.3106	14½	75½	.3113	13½	76½				.3119	12	78
.315-20	.3158	14¾	75¼	.3169	13¾	76¼	.3150	13	77	.3183	12¼	77¾
.320-25	.3210	15	75	.3226	14	76	.3209	13¼	76¾	.3247	12½	77½
.325-30	.3263	15¼	74¾	.3283	14¼	75¾	.3269	13½	76½			
.330-35	.3315	15½	74½	.3339	14½	75½	.3328	13¾	76¼	.3311	12¾	77¼
.335-40	.3367	15¾	74¼	.3395	14¾	75¼	.3387	14	76	.3375	13	77
.340-45	.3419	16	74				.3446	14¼	75¾	.3438	13¼	76¾
.345-50	.3471	16¼	73¾	.3451	15	75						
.350-55	.3523	16½	73½	.3508	15¼	74¾	.3506	14½	75½	.3502	13½	76½
.355-60	.3575	16¾	73¼	.3563	15½	74½	.3565	14¾	75¼	.3565	13¾	76¼
.360-65	.3627	17	73	.3620	15¾	74¼	.3624	15	75	.3629	14	76
.365-70	.3678	17¼	72¾	.3675	16	74	.3683	15¼	74¾	.3692	14¼	75¾
.370-75	.3730	17½	72½	.3731	16¼	73¾	.3742	15½	74½			
.375-80	.3781	17¾	72¼	.3787	16½	73½				.3756	14½	75½
.380-85	.3833	18	72	.3843	16¾	73¼	.3801	15¾	74¼	.3819	14¾	75¼
.385-90	.3884	18¼	71¾	.3898	17	73	.3859	16	74	.3882	15	75
.390-95	.3936	18½	71½				.3918	16¼	73¾	.3946	15¼	74¾
.395-00	.3987	18¾	71¼	.3954	17¼	72¾	.3977	16½	73½			
.400-05	.4038	19	71	.4011	17½	72½	.4035	16¾	73¼	.4009	15½	74½
.405-10	.4090	19¼	70¾	.4066	17¾	72¼	.4094	17	73	.4072	15¾	74¼
.410-15	.4140	19½	70½	.4121	18	72				.4135	16	74
.415-20	.4191	19¾	70¼	.4176	18¼	71¾	.4152	17¼	72¾	.4198	16¼	73¾
.420-25	.4242	20	70	.4231	18½	71½	.4210	17½	72½			
.425-30	.4293	20¼	69¾	.4287	18¾	71¼	.4268	17¾	72¼	.4260	16½	73½
.430-35	.4344	20½	69½	.4341	19	71	.4327	18	72	.4323	16¾	73¼
.435-40	.4394	20¾	69¼	.4396	19¼	70¾	.4384	18¼	71¾	.4385	17	73
.440-45	.4445	21	69				.4442	18½	71½	.4488	17¼	72¾
.445-50	.4495	21¼	68¾	.4451	19½	70½						

LEADS FROM .450 TO .600

Approximate Lead	Gear on Worm 24, 1st on Stud 86, 2nd on Stud 28, Gear on Screw 100	Angle to Set Spiral Head (Degrees)	Angle to Set Vertical Att. (Degrees)	Gear on Worm 24, 1st on Stud 86, 2nd on Stud 32, Gear on Screw 100	Angle to Set Spiral Head (Degrees)	Angle to Set Vertical Att. (Degrees)	Gear on Worm 24, 1st on Stud 72, 2nd on Stud 28, Gear on Screw 86	Angle to Set Spiral Head (Degrees)	Angle to Set Vertical Att. (Degrees)	Gear on Worm 24, 1st on Stud 72, 2nd on Stud 32, Gear on Screw 86	Angle to Set Spiral Head (Degrees)	Angle to Set Vertical Att. (Degrees)
.450-55	.4537	35½	54½	.4534	30½	59½	.4542	24¾	65¼	.4546	21½	68½
.455-60	.4593	36	54	.4566	30¾	59¼	.4585	25	65	.4596	21¾	68¼
.460-65	.4648	36½	53½	.4632	31¼	58¾	.4628	25¼	64¾	.4646	22	68
.465-70	.4675	36¾	53¼	.4666	31½	58½	.4671	25½	64½	.4696	22¼	67¾
.470-75	.4702	37	53	.4732	32	58	.4714	25¾	64¼	.4746	22½	67½
.475-80	.4756	37½	52½	.4798	32½	57½	.4756	26	64	.4796	22¾	67¼
.480-85	.4810	38	52	.4831	32¾	57¼	.4842	26½	63½	.4846	23	67
.485-90	.4890	38¾	51¼	.4864	33	57	.4884	26¾	63¼	.4896	23¼	66¾
.490-95	.4917	39	51	.4929	33½	56½	.4926	27	63	.4946	23½	66½
.495-00	.4970	39½	50½	.4994	34	56	.4968	27¼	62¾	.4995	23¾	66¼
.500-05	.5022	40	50	.5026	34¼	55¾	.5010	27½	62½	.5045	24	66
.505-10	.5075	40½	49½	.5058	34½	55½	.5052	27¾	62¼	.5094	24¼	65¾
.510-15	.5110	40¾	49¼	.5122	35	55	.5136	28¼	61¾	.5143	24½	65½
.515-20	.5178	41½	48½	.5186	35½	54½	.5177	28½	61½	.5193	24¾	65¼
.520-25	.5228	42	48	.5249	36	54	.5218	28¾	61¼	.5241	25	65
.525-30	.5279	42½	47½	.5280	36¼	53¾	.5260	29	61	.5290	25¼	64¾
.530-35	.5329	43	47	.5312	36½	53½	.5343	29½	60½	.5339	25½	64½
.535-40	.5378	43½	46½	.5374	37	53	.5384	29¾	60¼	.5389	25¾	64¼
.540-45	.5428	44	46	.5405	37¼	52¾	.5425	30	60	.5437	26	64
.545-50	.5452	44¼	45¾	.5497	38	52	.5466	30¼	59¾	.5486	26¼	63¾
.550-55	.5549	45¼	44¾	.5528	38¼	51¾	.5509	30½	59½	.5534	26½	63½
.555-60	.5573	45½	44½	.5559	38½	51½	.5588	31	59	.5582	26¾	63¼
.560-65	.5644	46¼	43¾	.5619	39	51	.5628	31¼	58¾	.5631	27	63
.565-70	.5667	46½	43½	.5650	39¼	50¾	.5669	31½	58½	.5679	27¼	62¾
.570-75	.5716	47	43	.5710	39¾	50¼	.5710	31¾	58¼	.5727	27½	62½
.575-80	.5761	47½	42½	.5770	40¼	49¾	.5750	32	58	.5775	27¾	62¼
.580-85	.5807	48	42	.5800	40½	49½	.5830	32½	57½	.5823	28	62
.585-90	.5852	48½	41½	.5888	41¼	48¾	.5869	32¾	57¼	.5871	28¼	61¾
.590-95	.5941	49½	40½	.5917	41½	48½	.5909	33	57	.5919	28½	61½
.595-00	.5986	50	40	.5975	42	48	.5988	33½	56½	.5966	28¾	61¼

Approximate Lead	Gear on Worm 24, 1st on Stud 72, 2nd on Stud 40, Gear on Screw 100	Angle to Set Spiral Head (Degrees)	Angle to Set Vertical Att. (Degrees)	Gear on Worm 24, 1st on Stud 48, 2nd on Stud 28, Gear on Screw 100	Angle to Set Spiral Head (Degrees)	Angle to Set Vertical Att. (Degrees)	Gear on Worm 24, 1st on Stud 64, 2nd on Stud 40, Gear on Screw 100	Angle to Set Spiral Head (Degrees)	Angle to Set Vertical Att. (Degrees)	Gear on Worm 28, 1st on Stud 56, 2nd on Stud 32, Gear on Screw 100	Angle to Set Spiral Head (Degrees)	Angle to Set Vertical Att. (Degrees)
.450-55	.4506	19¾	70¼	.4500	18¾	71¼	.4511	17½	72½	.4544	16½	73½
.455-60	.4561	20	70	.4558	19	71	.4573	17¾	72¼			
.460-65	.4615	20¼	69¾	.4616	19¼	70¾	.4635	18	72	.4611	16¾	73¼
.465-70	.4670	20½	69½	.4674	19½	70½	.4698	18¼	71¾	.4678	17	73
.470-75	.4724	20¾	69¼	.4731	19¾	70¼				.4745	17¼	72¾
.475-80	.4778	21	69	.4788	20	70	.4759	18½	71½			
.480-85	.4833	21¼	68¾	.4846	20¼	69¾	.4822	18¾	71¼	.4811	17½	72½
.485-90	.4887	21½	68½				.4883	19	71	.4878	17¾	72¼
.490-95	.4941	21¾	68¼	.4903	20½	69½	.4945	19¼	70¾	.4944	18	72
.495-00	.4995	22	68	.4960	20¾	69¼						
.500-05	.5049	22¼	67¾	.5018	21	69	.5007	19½	70½	.5010	18¼	71¾
.505-10				.5075	21¼	68¾	.5069	19¾	70¼	.5076	18½	71½
.510-15	.5103	22½	67½	.5131	21½	68½	.5130	20	70	.5143	18¾	71¼
.515-20	.5157	22¾	67¼	.5188	21¾	68¼	.5192	20¼	69¾			
.520-25	.5210	23	67	.5245	22	68				.5209	19	71
.525-30	.5264	23¼	66¾				.5253	20½	69½	.5275	19¼	70¾
.530-35	.5317	23½	66½	.5301	22¼	67¾	.5314	20¾	69¼	.5341	19½	70½
.535-40	.5371	23¾	66¼	.5358	22½	67½	.5375	21	69			
.540-45	.5424	24	66	.5414	22¾	67¼	.5436	21¼	68¾	.5407	19¾	70¼
.545-50	.5477	24¼	65¾	.5471	23	67	.5497	21½	68½	.5472	20	70
.550-55	.5530	24½	65½	.5527	23¼	66¾				.5538	20¼	69¾
.555-60	.5582	24¾	65¼	.5583	23½	66½	.5558	21¾	68¼			
.560-65	.5635	25	65	.5639	23¾	66¼	.5619	22	68	.5603	20½	69½
.565-70	.5688	25¼	64¾	.5695	24	66	.5680	22¼	67¾	.5668	20¾	69¼
.570-75	.5741	25½	64½				.5740	22½	67½	.5734	21	69
.575-80	.5793	25¾	64¼	.5751	24¼	65¾				.5799	21¼	68¾
.580-85	.5846	26	64	.5806	24½	65½	.5800	22¾	67¼			
.585-90	.5898	26¼	63¾	.5862	24¾	65¼	.5861	23	67	.5864	21½	68½
.590-95	.5949	26½	63½	.5917	25	65	.5921	23¼	66¾	.5928	21¾	68¼
.595-00				.5972	25¼	64¾	.5981	23½	66½	.5994	22	68

Approximate Lead	Gear on Worm 24, 1st on Stud 86, 2nd on Stud 32, Gear on Screw 100	Angle to Set Spiral Head (Degrees)	Angle to Set Vertical Att. (Degrees)	Gear on Worm 24, 1st on Stud 72, 2nd on Stud 28, Gear on Screw 86	Angle to Set Spiral Head (Degrees)	Angle to Set Vertical Att. (Degrees)	Gear on Worm 24, 1st on Stud 72, 2nd on Stud 32, Gear on Screw 86	Angle to Set Spiral Head (Degrees)	Angle to Set Vertical Att. (Degrees)	Gear on Worm 24, 1st on Stud 72, 2nd on Stud 40, Gear on Screw 100	Angle to Set Spiral Head (Degrees)	Angle to Set Vertical Att. (Degrees)	Gear on Worm 24, 1st on Stud 48, 2nd on Stud 28, Gear on Screw 100	Angle to Set Spiral Head (Degrees)	Angle to Set Vertical Att. (Degrees)	Gear on Worm 24, 1st on Stud 64, 2nd on Stud 40, Gear on Screw 100	Angle to Set Spiral Head (Degrees)	Angle to Set Vertical Att. (Degrees)	Gear on Worm 28, 1st on Stud 56, 2nd on Stud 32, Gear on Screw 100	Angle to Set Spiral Head (Degrees)	Angle to Set Vertical Att. (Degrees)	Gear on Worm 24, 1st on Stud 72, 2nd on Stud 44, Gear on Screw 86	Angle to Set Spiral Head (Degrees)	Angle to Set Vertical Att. (Degrees)
.600-05	.6033	42½	47½	.6028	33¾	56¼	.6013	29	61	.6002	26¾	63¼	.6028	25½	64½	.6041	23¾	66¼				.6042	20¾	69¼
.605-10	.6090	43	47	.6067	34	56	.6060	29¼	60¾	.6054	27	63	.6083	25¾	64¼									
.610-15	.6124	43¼	46¾	.6146	34½	55½	.6108	29½	60½	.6106	27¼	62¾	.6138	26	64	.6101	24	66	.6123	22½	67½	.6112	21	69
.615-20	.6175	43¾	46¼	.6184	34¾	55¼	.6155	29¾	60¼	.6158	27½	62½	.6192	26¼	63¾	.6160	24¼	65¾	.6187	22¾	67¼	.6182	21¼	68¾
.620-25	.6231	44¼	45¾	.6223	35	55	.6202	30	60	.6209	27¾	62¼	.6247	26½	63½	.6220	24½	65½						
.625-30	.6286	44¾	45¼	.6262	35¼	54¾	.6295	30½	59½	.6260	28	62				.6279	24¾	65¼	.6251	23	67	.6251	21½	68½
.630-35	.6314	45	45	.6300	35½	54½	.6342	30¾	59¼	.6312	28¼	61¾	.6302	26¾	63¼	.6339	25	65	.6316	23¼	66¾	.6320	21¾	68¼
.635-40	.6369	45½	44½	.6377	36	54	.6388	31	59	.6363	28½	61½	.6356	27	63	.6398	25¼	64¾	.6380	23½	66½	.6389	22	68
.640-45	.6423	46	44				.6434	31¼	58¾	.6414	28¾	61¼	.6410	27¼	62¾				.6444	23¾	66¼			
.645-50	.6477	46½	43½	.6491	36¾	53¼	.6481	31½	58½	.6464	29	61	.6465	27½	62½	.6458	25½	64½				.6457	22¼	67¾
.650-55	.6504	46¾	43¼	.6530	37	53	.6526	31¾	58¼	.6516	29¼	60¾	.6519	27¾	62¼	.6516	25¾	64¼	.6508	24	66	.6526	22½	67½
.655-60	.6584	47½	42½	.6567	37¼	52¾	.6572	32	58	.6566	29½	60½	.6573	28	62	.6576	26	64	.6571	24¼	65¾	.6595	22¾	67¼
.660-65	.6637	48	42	.6605	37½	52½	.6618	32¼	57¾	.6617	29¾	60¼	.6627	28¼	61¾	.6634	26¼	63¾	.6634	24½	65½			
.665-70	.6688	48½	41½	.6680	38	52	.6664	32½	57½	.6667	30	60	.6680	28½	61½	.6693	26½	63½	.6698	24¾	65¼	.6664	23	67
.670-75	.6739	49	41	.6717	38¼	51¾	.6710	32¾	57¼	.6718	30¼	59¾	.6734	28¾	61¼							.6732	23¼	66¾
.675-80	.6790	49½	40½	.6754	38½	51½	.6755	33	57	.6767	30½	59¾	.6788	29	61	.6751	26¾	63¼	.6762	25	65			
.680-85	.6840	50	40	.6828	39	51	.6845	33½	56½	.6818	30¾	59¼	.6841	29¼	60¾	.6810	27	63	.6825	25¼	64¾	.6801	23½	66½
.685-90	.6890	50½	39½	.6865	39¼	50¾	.6890	33¾	56¼	.6868	31	59	.6894	29½	60½	.6868	27¼	62¾	.6888	25½	64½	.6868	23¾	66¼
.690-95	.6940	51	39	.6901	39½	50½	.6935	34	56	.6918	31¼	58¾	.6948	29¾	60¼	.6926	27½	62½				.6937	24	66
.695-00	.6988	51½	38½	.6974	40	50	.6980	34¼	55¾	.6967	31½	58½				.6984	27¾	62¼	.6951	25¾	64¼			
.700-05	.7036	52	38	.7046	40½	49½	.7025	34½	55½	.7017	31¾	58¼	.7000	30	60	.7042	28	62	.7014	26	64	.7004	24¼	65¾
.705-10	.7085	52½	37½	.7082	40¾	49¼	.7070	34¾	55¼	.7066	32	58	.7053	30¼	59¾				.7076	26¼	63¾	.7072	24½	65½
.710-15	.7132	53	37	.7118	41	49	.7114	35	55	.7116	32¼	57¾	.7106	30½	59½	.7100	28¼	61¾	.7140	26½	63½	.7140	24¾	65¼
.715-20	.7178	53½	36½	.7189	41½	48½	.7158	35¼	54¾	.7165	32½	57½	.7159	30¾	59¼	.7157	28½	61½						
.720-25	.7224	54	36	.7225	41¾	48¼	.7247	35¾	54¼	.7214	32¾	57¼	.7211	31	59	.7214	28¾	61¼	.7202	26¾	63¼	.7208	25	65
.725-30	.7270	54½	35½	.7260	42	48	.7291	36	54	.7263	33	57	.7264	31¼	58¾	.7272	29	61	.7264	27	63	.7275	25¼	64¾
.730-35	.7315	55	35	.7330	42½	47½	.7334	36¼	53¾	.7312	33¼	56¾	.7315	31½	58½	.7329	29¼	60¾	.7326	27¼	62¾	.7342	25½	64½
.735-40	.7360	55½	34½	.7397	43	47	.7378	36½	53½	.7360	33½	56½	.7367	31¾	58¼	.7386	29½	60½	.7387	27½	62½			
.740-45	.7404	56	34	.7441	43¼	46¾	.7421	36¾	53¼	.7409	33¾	56¼	.7419	32	58	.7443	29¾	60¼	.7449	27¾	62¼	.7409	25¾	64¼
.745-50	.7490	57	33	.7467	43½	46½	.7464	37	53	.7456	34	56	.7471	32¼	57¾							.7476	26	64

LEADS FROM .750 TO .900

Approximate Lead	Gear on Worm 24, 1st on Stud 72, 2nd on Stud 28, Gear on Screw 86	Angle to Set Spiral Head (Degrees)	Angle to Set Vertical Att. (Degrees)	Gear on Worm 24, 1st on Stud 72, 2nd on Stud 32, Gear on Screw 86	Angle to Set Spiral Head (Degrees)	Angle to Set Vertical Att. (Degrees)	Gear on Worm 24, 1st on Stud 72, 2nd on Stud 40, Gear on Screw 100	Angle to Set Spiral Head (Degrees)	Angle to Set Vertical Att. (Degrees)	Gear on Worm 24, 1st on Stud 48, 2nd on Stud 28, Gear on Screw 100	Angle to Set Spiral Head (Degrees)	Angle to Set Vertical Att. (Degrees)	Gear on Worm 24, 1st on Stud 64, 2nd on Stud 40, Gear on Screw 100	Angle to Set Spiral Head (Degrees)	Angle to Set Vertical Att. (Degrees)	Gear on Worm 28, 1st on Stud 56, 2nd on Stud 32, Gear on Screw 100	Angle to Set Spiral Head (Degrees)	Angle to Set Vertical Att. (Degrees)	Gear on Worm 24, 1st on Stud 72, 2nd on Stud 44, Gear on Screw 86	Angle to Set Spiral Head (Degrees)	Angle to Set Vertical Att. (Degrees)	Gear on Worm 24, 1st on Stud 64, 2nd on Stud 48, Gear on Screw 100	Angle to Set Spiral Head (Degrees)	Angle to Set Vertical Att. (Degrees)
.750-55	.7537	44	46	.7508	37¼	52¾	.7505	34¼	55¾	.7522	32½	57½	.7500	30	60	.7511	28	62	.7543	26¼	63¾	.7535	24¾	65¼
.755-60	.7570	44¼	45¾	.7550	37½	52½	.7553	34½	55½	.7573	32¾	57¼	.7556	30¼	59¾	.7573	28¼	61¾						
.760-65	.7605	44½	45½	.7636	38	52	.7648	35	55	.7626	33	57	.7613	30½	59½	.7634	28½	61½	.7610	26½	63½	.7607	25	65
.765-70	.7672	45	45	.7679	38¼	51¾	.7696	35¼	54¾	.7677	33¼	56¾	.7670	30¾	59¼	.7695	28¾	61¼	.7677	26¾	63¼	.7678	25¼	64¾
.770-75	.7738	45½	44½	.7721	38½	51½	.7744	35½	54½	.7728	33½	56½	.7725	31	59				.7743	27	63	.7749	25½	64½
.775-80	.7772	45¾	44¼	.7763	38¾	51¼	.7790	35¾	54¼	.7778	33¾	56¼	.7781	31¼	58¾	.7756	29	61						
.780-85	.7805	46	44	.7805	39	51	.7838	36	54	.7829	34	56	.7837	31½	58½	.7817	29¼	60¾	.7808	27¼	62¾	.7820	25¾	64¼
.785-90	.7869	46½	43½	.7889	39½	50½	.7884	36¼	53¾	.7880	34¼	55¾	.7892	31¾	58¼	.7878	29½	60½	.7875	27½	62½	.7890	26	64
.790-95	.7935	47	43	.7930	39¾	50¼	.7932	36½	53½	.7930	34½	55½	.7948	32	58	.7939	29¾	60¼	.7941	27¾	62¼			
.795-00	.7999	47½	42½	.7973	40	50	.7978	36¾	53¼	.7980	34¾	55¼										.7961	26¼	63¾
.800-05	.8030	47¾	42¼	.8014	40¼	49¾	.8025	37	53	.8030	35	55	.8004	32¼	57¾	.8000	30	60	.8007	28	62	.8032	26½	63½
.805-10	.8064	48	42	.8055	40½	49½	.8072	37¼	52¾	.8080	35¼	54¾	.8060	32½	57½	.8060	30¼	59¾	.8073	28¼	61¾			
.810-15	.8126	48½	41½	.8137	41	49	.8117	37½	52½	.8131	35½	54½	.8114	32¾	57¼	.8120	30½	59½	.8138	28½	61½	.8102	26¾	63¼
.815-20	.8188	49	41	.8178	41¼	48¾	.8164	37¾	52¼	.8180	35¾	54¼	.8170	33	57	.8180	30¾	59¼				.8172	27	63
.820-25	.8219	49¼	40¾	.8128	41½	48½	.8210	38	52	.8230	36	54	.8224	33¼	56¾	.8240	31	59	.8203	28¾	61¼	.8241	27¼	62¾
.825-30	.8250	49½	40½	.8299	42	48	.8255	38¼	51¾	.8278	36¼	53¾	.8279	33½	56½				.8268	29	61			
.830-35	.8311	50	40	.8339	42¼	47¾	.8301	38½	51½	.8328	36½	53½	.8334	33¾	56¼	.8300	31¼	58¾	.8333	29¼	60¾	.8311	27½	62½
.835-40	.8371	50½	39½	.8379	42½	47½	.8391	39	51	.8378	36¾	53¼	.8388	34	56	.8359	31½	58½	.8398	29½	60½	.8380	27¾	62¼
.840-45	.8431	51	39	.8419	42¾	47¼	.8437	39¼	50¾	.8426	37	53	.8442	34¼	55¾	.8419	31¾	58¼						
.845-50	.8491	51½	38½	.8459	43	47	.8482	39½	50½	.8474	37¼	52¾	.8496	34½	55½	.8478	32	58	.8463	29¾	60¼	.8450	28	62
.850-55	.8520	51¾	38¼	.8538	43½	46½	.8527	39¾	50¼	.8523	37½	52½				.8538	32¼	57¾	.8527	30	60	.8520	28¼	61¾
.855-60	.8550	52	38	.8577	43¾	46¼	.8572	40	50	.8572	37¾	52¼	.8550	34¾	55¼	.8596	32½	57½	.8592	30¼	59¾	.8589	28½	61½
.860-65	.8608	52½	37½	.8616	44	46	.8616	40¼	49¾	.8620	38	52	.8604	35	55									
.865-70	.8665	53	37	.8693	44½	45½	.8660	40½	49½	.8668	38¼	51¾	.8657	35¼	54¾	.8655	32¾	57¼	.8656	30½	59½	.8657	28¾	61¼
.870-75	.8722	53½	36½	.8732	44¾	45¼	.8784	41	49	.8716	38½	51½	.8710	35½	54½	.8714	33	57	.8720	30¾	59¼	.8726	29	61
.875-80	.8778	54	36	.8770	45	45	.8792	41¼	48¾	.8763	38¾	51¼	.8763	35¾	54¾	.8772	33¼	56¾	.8784	31	59	.8794	29¼	60¾
.880-85	.8832	54½	35½	.8846	45½	44½	.8835	41½	48½	.8810	39	51	.8816	36	54	.8830	33½	56½	.8847	31¼	58¾			
.885-90	.8887	55	35	.8884	45¾	44¼	.8880	41¾	48¼	.8858	39¼	50¾	.8869	36¼	53¾	.8888	33¾	56¼				.8863	29½	60½
.890-95	.8942	55½	34½	.8922	46	44	.8923	42	48	.8906	39½	50½	.8922	36½	53½	.8947	34	56	.8911	31½	58½	.8931	29¾	60¼
.895-00	.8995	56	34	.8996	46½	43½	.8966	42¼	47¾	.8953	39¾	50¼	.8974	36¾	53¼				.8975	31¾	58¼			

Approximate Lead	Gear on Worm 24, 1st on Stud 72, 2nd on Stud 28, Gear on Screw 86	Angle to Set Spiral Head (Degrees)	Angle to Set Vertical Att. (Degrees)	Gear on Worm 24, 1st on Stud 72, 2nd on Stud 32, Gear on Screw 86	Angle to Set Spiral Head (Degrees)	Angle to Set Vertical Att. (Degrees)	Gear on Worm 24, 1st on Stud 72, 2nd on Stud 40, Gear on Screw 100	Angle to Set Spiral Head (Degrees)	Angle to Set Vertical Att. (Degrees)	Gear on Worm 24, 1st on Stud 48, 2nd on Stud 28, Gear on Screw 100	Angle to Set Spiral Head (Degrees)	Angle to Set Vertical Att. (Degrees)
.900-05	.9047	56½	33½	.9034	46¾	43¼	.9009	42½	46½	.9000	40	50
.905-10	.9074	56¾	33¼	.9071	47	43	.9094	43	47	.9093	40½	49½
.910-15	.9100	57	33	.9145	47½	42½	.9146	43¼	46¾	.9139	40¾	49¼
.915-20	.9151	57½	32½	.9181	47¾	42¼	.9178	43½	46½	.9186	41	49
.920-25	.9200	58	32	.9217	48	42	.9221	43¾	46¼	.9231	41¼	48¾
.925-30	.9250	58½	31½	.9290	48½	41½	.9263	44	46	.9277	41½	48½
.930-35	.9300	59	31	.9325	48¾	41¼	.9346	44½	45½	.9323	41¾	48¼
.935-40	.9395	60	30	.9360	49	41	.9388	44¾	45¼	.9368	42	48
.940-45	.9443	60½	29½	.9431	49½	40½	.9429	45	45	.9414	42¼	47¾
.945-50	.9489	61	29	.9466	49¾	40¼	.9471	45¼	44¾	.9458	42½	47½
.950-55	.9535	61½	28½	.9501	50	40	.9511	45½	44½	.9548	43	47
.955-60	.9580	62	28	.9570	50½	39½	.9591	46	44	.9593	43¼	46¾
.960-65	.9624	62½	27½	.9640	51	39	.9633	46¼	43¾	.9638	43½	46½
.965-70	.9667	63	27	.9673	51¼	38¾	.9672	46½	43½	.9682	43¾	46¼
.970-75	.9712	63½	26½	.9707	51½	38½	.9712	46¾	43¼	.9726	44	46
.975-80	.9754	64	26	.9774	52	38	.9752	47	43	.9770	44¼	45¾
.980-85	.9836	65	25	.9840	52½	37½	.9831	47½	42½	.9813	44½	45½
.985-90	.9875	65½	24½	.9863	52¾	37¼	.9870	47¾	42¼	.9856	44¾	45¼
.990-95	.9914	66	24	.9906	53	37	.9911	48	42	.9900	45	45
.995-00	.9952	66½	23½	.9970	53½	36½	.9987	48½	41½	.9986	45½	44½
1.000-05	1.0027	67½	22½	1.0035	54	36	1.0026	48¾	41¼	1.0028	45¾	44¼
1.005-10	1.0062	68	22	1.0098	54½	35½	1.0064	49	41	1.0071	46	44
1.010-15	1.0132	69	21	1.0130	54¾	35¼	1.0140	49½	40½	1.0114	46¼	43¾
1.015-20	1.0198	70	20	1.0160	55	35	1.0178	49¾	40¼	1.0156	46½	43½
1.020-25	1.0230	70½	19½	1.0222	55½	34½	1.0215	50	40	1.0240	47	43
1.025-30	1.0261	71	19	1.0252	55¾	34¼	1.0288	50½	39½	1.0282	47¼	42¾
1.030-35	1.0322	72	18	1.0342	56½	33½	1.0326	50¾	39¼	1.0323	47½	42½
1.035-40	1.0350	72½	17½	1.0371	56¾	33¼	1.0362	51	39	1.0364	47¾	42¼
1.040-45	1.0432	74	16	1.0404	57	33	1.0435	51½	38½	1.0405	48	42
1.045-50	1.0482	75	15	1.0461	57½	32½	1.0471	51¾	38¼	1.0486	48½	41½

Approximate Lead	Gear on Worm 24, 1st on Stud 64, 2nd on Stud 40, Gear on Screw 100	Angle to Set Spiral Head (Degrees)	Angle to Set Vertical Att. (Degrees)	Gear on Worm 28, 1st on Stud 56, 2nd on Stud 32, Gear on Screw 100	Angle to Set Spiral Head (Degrees)	Angle to Set Vertical Att. (Degrees)	Gear on Worm 24, 1st on Stud 72, 2nd on Stud 44, Gear on Screw 86	Angle to Set Spiral Head (Degrees)	Angle to Set Vertical Att. (Degrees)	Gear on Worm 24, 1st on Stud 64, 2nd on Stud 48, Gear on Screw 100	Angle to Set Spiral Head (Degrees)	Angle to Set Vertical Att. (Degrees)
.900-05	.9027	37	53	.9004	34¼	55¾	.9037	32	58	.9000	30	60
.905-10	.9079	37¼	52¾	.9062	34½	55½				.9068	30¼	59¾
.910-15	.9131	37½	52½	.9120	34¾	55¼	.9101	32¼	57¾	.9136	30½	59½
.915-20	.9183	37¾	52¼	.9177	35	55	.9164	32½	57½			
.920-25	.9234	38	52	.9234	35¼	54¾	.9226	32¾	57¼	.9204	30¾	59¼
.925-30	.9286	38¼	51¾	.9291	35½	54½	.9288	33	57	.9270	31	59
.930-35	.9337	38½	51½	.9347	35¾	54¼				.9337	31¼	58¾
.935-40	.9388	38¾	51¼				.9351	33¼	56¾			
.940-45	.9439	39	51	.9404	36	54	.9413	33½	56½	.9404	31½	58½
.945-50	.9490	39¼	50¾	.9460	36¼	53¾	.9475	33¾	56¼	.9471	31¾	58¼
.950-55	.9541	39½	50½	.9516	36½	53½	.9537	34	56	.9538	32	58
.955-60	.9592	39¾	50¼	.9572	36¾	53¼	.9598	34¼	55¾			
.960-65	.9642	40	50	.9629	37	53				.9605	32¼	57¾
.965-70	.9691	40¼	49¾	.9685	37¼	52¾	.9661	34½	55½	.9672	32½	57½
.970-75	.9741	40½	49½	.9740	37½	52½	.9722	34¾	55¼	.9737	32¾	57¼
.975-80	.9791	40¾	49¼	.9795	37¾	52¼	.9783	35	55			
.980-85	.9840	41	49				.9843	35¼	54¾	.9803	33	57
.985-90	.9890	41¼	48¾	.9850	38	52				.9869	33¼	56¾
.990-95	.9938	41½	48½	.9905	38¼	51¾	.9904	35½	54½	.9934	33½	56½
.995-00	.9988	41¾	48¼	.9959	38½	51½	.9964	35¾	54¼			
1.000-05	1.0037	42	48	1.0014	38¾	51¼	1.0024	36	54	1.0000	33¾	56¼
1.005-10	1.0085	42¼	47¾	1.0068	39	51	1.0083	36¼	53¾	1.0065	34	56
1.010-15	1.0135	42½	47½	1.0123	39¼	50¾	1.0144	36½	53½	1.0130	34¼	55¾
1.015-20	1.0183	42¾	47¼	1.0177	39½	50½				1.0195	34½	55½
1.020-25	1.0230	43	47	1.0231	39¾	50¼	1.0203	36¾	53¼			
1.025-30	1.0287	43¼	46¾	1.0284	40	50	1.0263	37	53	1.0260	34¾	55¼
1.030-35	1.0323	43½	46½	1.0337	40¼	49¾	1.0323	37¼	52¾	1.0324	35	55
1.035-40	1.0371	43¾	46¼	1.0391	40½	49½	1.0382	37½	52½	1.0388	35¼	54¾
1.040-45	1.0419	44	46	1.0443	40¾	49¼	1.0440	37¾	52¼			
1.045-50	1.0466	44¼	45¾	1.0496	41	49				1.0452	35½	54½

LEADS FROM 1.050 TO 1.200

Approximate Lead	Gear on Worm 24, 1st on Stud 72, 2nd on Stud 32, Gear on Screw 86	Angle to Set Spiral Head (Degrees)	Angle to Set Vertical Att. (Degrees)	Gear on Worm 24, 1st on Stud 72, 2nd on Stud 40, Gear on Screw 100	Angle to Set Spiral Head (Degrees)	Angle to Set Vertical Att. (Degrees)	Gear on Worm 24, 1st on Stud 48, 2nd on Stud 28, Gear on Screw 100	Angle to Set Spiral Head (Degrees)	Angle to Set Vertical Att. (Degrees)	Gear on Worm 24, 1st on Stud 64, 2nd on Stud 40, Gear on Screw 100	Angle to Set Spiral Head (Degrees)	Angle to Set Vertical Att. (Degrees)	Gear on Worm 28, 1st on Stud 56, 2nd on Stud 32, Gear on Screw 100	Angle to Set Spiral Head (Degrees)	Angle to Set Vertical Att. (Degrees)	Gear on Worm 24, 1st on Stud 72, 2nd on Stud 44, Gear on Screw 86	Angle to Set Spiral Head (Degrees)	Angle to Set Vertical Att. (Degrees)	Gear on Worm 24, 1st on Stud 64, 2nd on Stud 48, Gear on Screw 100	Angle to Set Spiral Head (Degrees)	Angle to Set Vertical Att. (Degrees)	Gear on Worm 24, 1st on Stud 40, 2nd on Stud 32, Gear on Screw 100	Angle to Set Spiral Head (Degrees)	Angle to Set Vertical Att. (Degrees)
1.050-55	1.0518	58	32	1.0508	52	38	1.0527	48¾	41¼	1.0513	44½	45½	1.0549	41¼	48¾	1.0500	38	52	1.0515	35¾	54¼	1.0527	33¼	56¾
1.055-60	1.0575	58½	31½	1.0577	52½	37½	1.0567	49	41	1.0559	44¾	45¼				1.0558	38¼	51¾	1.0579	36	54	1.0597	33½	56½
1.060-65	1.0632	59	31	1.0613	52¾	37¼	1.0646	49½	40½	1.0606	45	45	1.0601	41½	48½	1.0617	38½	51½	1.0642	36¼	53¾			
1.065-70	1.0685	59½	30½	1.0650	53	37	1.0685	49¾	40¼	1.0698	45½	44½	1.0653	41¾	48¼	1.0674	38¾	51¼				1.0667	33¾	56¼
1.070-75	1.0740	60	30	1.0720	53½	36½	1.0725	50	40	1.0744	45¾	44¼	1.0706	42	48	1.0732	39	51	1.0706	36½	53½	1.0736	34	56
1.075-80	1.0795	60½	29½	1.0788	54	36	1.0764	50¼	39¾	1.0790	46	44	1.0758	42¼	47¾	1.0790	39¼	50¾	1.0769	36¾	53¼			
1.080-85	1.0848	61	29	1.0822	54¼	35¾	1.0803	50½	39½	1.0835	46¼	43¾	1.0809	42½	47½	1.0848	39½	50½	1.0832	37	53	1.0805	34¼	55¾
1.085-90	1.0871	61¼	28¾	1.0856	54½	35½	1.0880	51	39	1.0879	46½	43½	1.0861	42¾	47¼				1.0895	37¼	52¾	1.0874	34½	55½
1.090-95	1.0900	61½	28½	1.0923	55	35	1.0919	51¼	38¾	1.0924	46¾	43¼	1.0911	43	47	1.0906	39¾	50¼				1.0943	34¾	55¼
1.095-00	1.0950	62	28	1.0990	55½	34½	1.0958	51½	38½	1.0969	47	43	1.0963	43¼	46¾	1.0962	40	50	1.0957	37½	52½			
1.100-05	1.1000	62½	27½	1.1022	55¾	34¼	1.1033	52	38	1.1014	47¼	42¾	1.1014	43½	46½	1.1019	40¼	49¾	1.1019	37¾	52¼	1.1012	35	55
1.105-10	1.1050	63	27	1.1055	56	34	1.1070	52¼	37¾	1.1058	47½	42½	1.1063	43¾	46¼	1.1076	40½	49½	1.1081	38	52	1.1080	35¼	54¾
1.110-15	1.1148	64	26	1.1119	56½	33½	1.1107	52½	37½	1.1148	48	42	1.1115	44	46	1.1132	40¾	49¼	1.1142	38¼	51¾	1.1149	35½	54½
1.115-20	1.1195	64½	25½	1.1184	57	33	1.1181	53	37	1.1190	48¼	41¾	1.1163	44¼	45¾	1.1188	41	49						
1.120-25	1.1240	65	25	1.1247	57½	32½	1.1218	53¼	36¾	1.1234	48½	41½	1.1213	44½	45½	1.1244	41¼	48¾	1.1204	38½	51½	1.1217	35¾	54¼
1.125-30	1.1286	65½	24½	1.1278	57¾	32¼	1.1254	53½	36½	1.1277	48¾	41¼	1.1262	44¾	45¼				1.1266	38¾	51¼	1.1285	36	54
1.130-35	1.1330	66	24	1.1308	58	32	1.1327	54	36	1.1320	49	41	1.1313	45	45	1.1300	41½	48½	1.1327	39	51			
1.135-40	1.1375	66½	23½	1.1369	58½	31½	1.1398	54½	35½	1.1363	49¼	40¾	1.1362	45¼	44¾	1.1357	41¾	48¼	1.1388	39¼	50¾	1.1352	36¼	53¾
1.140-45	1.1417	67	23	1.1430	59	31	1.1433	54¾	35¼	1.1405	49½	40½	1.1411	45½	44½	1.1412	42	48	1.1449	39½	50½	1.1420	36½	53½
1.145-50	1.1459	67½	22½	1.1489	59½	30½	1.1468	55	35	1.1490	50	40	1.1460	45¾	44¼	1.1467	42¼	47¾				1.1487	36¾	53¼
1.150-55	1.1500	68	22	1.1548	60	30	1.1538	55½	34½	1.1532	50¼	39¾	1.1508	46	44	1.1522	42½	47½	1.1510	39¾	50¼			
1.155-60	1.1598	69¼	20¾	1.1578	60¼	29¾	1.1572	55¾	34¼	1.1573	50½	39½	1.1557	46¼	43¾	1.1578	42¾	47¼	1.1570	40	50	1.1554	37	53
1.160-65	1.1618	69½	20½	1.1606	60½	29½	1.1607	56	34	1.1615	50¾	39¼	1.1606	46½	43½	1.1631	43	47	1.1629	40¼	49¾	1.1621	37¼	52¾
1.165-70	1.1655	70	20	1.1662	61	29	1.1674	56½	33½	1.1656	51	39	1.1654	46¾	43¼	1.1685	43¼	46¾	1.1689	40½	49½	1.1687	37½	52½
1.170-75	1.1727	71	19	1.1718	61½	28½	1.1742	57	33	1.1738	51½	38½	1.1700	47	43	1.1738	43½	46½	1.1749	40¾	49¼			
1.175-80	1.1761	71½	18½	1.1775	62	28	1.1774	57¼	32¾	1.1778	51¾	38¼	1.1795	47½	42½	1.1792	43¾	46¼				1.1753	37¾	52¼
1.180-85	1.1829	72½	17½	1.1828	62½	27	1.1808	57½	32½	1.1819	52	38	1.1841	47¾	42¼	1.1847	44	46	1.1808	41	49	1.1820	38	52
1.185-90	1.1860	73	17	1.1881	63	27	1.1871	58	32	1.1861	52¼	37¾	1.1891	48	42				1.1867	41¼	48¾	1.1887	38¼	51¾
1.190-95	1.1922	74	16	1.1934	63½	26½	1.1936	58½	31½	1.1900	52½	37½	1.1935	48¼	41¾	1.1900	44¼	45¾	1.1926	41½	48½			
1.195-00	1.1980	75	15	1.1986	64	26	1.1968	58¾	31¼	1.1980	53	37	1.1981	48½	41½	1.1953	44½	45½	1.1985	41¾	48¼	1.1951	38½	51½

Approximate Lead	Gear on Worm 24, 1st on Stud 48, 2nd on Stud 28, Gear on Screw 100	Angle to Set Spiral Head (Degrees)	Angle to Set Vertical Att. (Degrees)	Gear on Worm 24, 1st on Stud 64, 2nd on Stud 40, Gear on Screw 100	Angle to Set Spiral Head (Degrees)	Angle to Set Vertical Att. (Degrees)	Gear on Worm 28, 1st on Stud 56, 2nd on Stud 32, Gear on Screw 100	Angle to Set Spiral Head (Degrees)	Angle to Set Vertical Att. (Degrees)	Gear on Worm 24, 1st on Stud 72, 2nd on Stud 44, Gear on Screw 86	Angle to Set Spiral Head (Degrees)	Angle to Set Vertical Att. (Degrees)
1.200-05	1.2000	59	31	1.2018	53¼	36¾	1.2029	48¾	41¼	1.2007	44¾	45¼
1.205-10	1.2062	59½	30½	1.2058	53½	36½	1.2075	49	41	1.2060	45	45
1.210-15	1.2124	60	30	1.2135	54	36	1.2120	49¼	40¾	1.2112	45¼	44¾
1.215-20	1.2185	60½	29½	1.2172	54¼	35¾	1.2165	49½	40½	1.2163	45½	44½
1.220-25	1.2245	61	29	1.2210	54½	35½	1.2211	49¾	40¼	1.2216	45¾	44¼
1.225-30	1.2274	61¼	28¾	1.2287	55	35	1.2256	50	40	1.2268	46	44
1.230-35	1.2302	61½	28½	1.2324	55¼	34¾	1.2345	50½	39½	1.2319	46¼	43¾
1.235-40	1.2362	62	28	1.2362	55½	34½	1.2389	50¾	39¼	1.2370	46½	43½
1.240-45	1.2418	62½	27½	1.2435	56	34	1.2435	51	39	1.2421	46¾	43¼
1.245-50	1.2474	63	27	1.2470	56¼	33¾	1.2474	51¼	38¾	1.2472	47	43
1.250-55	1.2529	63½	26½	1.2506	56½	33½	1.2521	51½	38½	1.2523	47¼	42¾
1.255-60	1.2584	64	26	1.2580	57	33	1.2563	51¾	38¼	1.2573	47½	42½
1.260-65	1.2636	64½	25½	1.2614	57¼	32¾	1.2607	52	38	1.2623	47¾	42¼
1.265-70	1.2689	65	25	1.2655	57½	32½	1.2694	52½	37½	1.2675	48	42
1.270-75	1.2739	65½	24½	1.2718	58	32	1.2736	52¾	37¼	1.2722	48¼	41¾
1.275-80	1.2790	66	24	1.2788	58½	31½	1.2778	53	37	1.2772	48½	41½
1.280-85	1.2840	66½	23½	1.2821	58¾	31¼	1.2820	53¼	36¾	1.2822	48¾	41¼
1.285-90	1.2887	67	23	1.2858	59	31	1.2861	53½	36½	1.2870	49	41
1.290-95	1.2934	67½	22½	1.2923	59½	30½	1.2902	53¾	36¼	1.2920	49¼	40¾
1.295-00	1.2980	68	22	1.2989	60	30	1.2943	54	36	1.2968	49½	40½
1.300-05	1.3025	68½	21½	1.3021	60¼	29¾	1.3024	54½	35½	1.3017	49¾	40¼
1.305-10	1.3069	69	21	1.3055	60½	29½	1.3065	54¾	35¼	1.3063	50	40
1.310-15	1.3112	69½	20½	1.3118	61	29	1.3105	55	35	1.3111	50¼	39¾
1.315-20	1.3155	70	20	1.3180	61½	28½	1.3185	55½	34½	1.3160	50½	39½
1.320-25	1.3237	71	19	1.3244	62	28	1.3224	55¾	34¼	1.3207	50¾	39¼
1.325-30	1.3276	71½	18½	1.3275	62¼	27¾	1.3262	56	34	1.3252	51	39
1.330-35	1.3315	72	18	1.3304	62½	27½	1.3340	56½	33½	1.3346	51½	38½
1.335-40	1.3389	73	17	1.3363	63	27	1.3384	56¾	33¼	1.3391	51¾	38¼
1.340-45	1.3424	73½	16½	1.3423	63½	26½	1.3419	57	33	1.3439	52	38
1.345-50	1.3458	74	16	1.3481	64	26	1.3493	57½	32½	1.3487	52¼	37¾

Approximate Lead	Gear on Worm 24, 1st on Stud 64, 2nd on Stud 48, Gear on Screw 100	Angle to Set Spiral Head (Degrees)	Angle to Set Vertical Att. (Degrees)	Gear on Worm 24, 1st on Stud 40, 2nd on Stud 32, Gear on Screw 100	Angle to Set Spiral Head (Degrees)	Angle to Set Vertical Att. (Degrees)	Gear on Worm 28, 1st on Stud 64, 2nd on Stud 40, Gear on Screw 86	Angle to Set Spiral Head (Degrees)	Angle to Set Vertical Att. (Degrees)	Gear on Worm 24, 1st on Stud 72, 2nd on Stud 56, Gear on Screw 86	Angle to Set Spiral Head (Degrees)	Angle to Set Vertical Att. (Degrees)
1.200-05	1.2043	42	48	1.2017	38¾	51¼	1.2032	36¼	53¾			
1.205-10				1.2082	39	51				1.2059	33¾	56¼
1.210-15	1.2102	42¼	47¾	1.2149	39¼	50¾	1.2104	36½	53½	1.2138	34	56
1.215-20	1.2160	42½	47½				1.2175	36¾	53¼			
1.220-25	1.2219	42¾	47¼	1.2212	39½	50½	1.2246	37	53	1.2216	34¼	55¾
1.225-30	1.2275	43	47	1.2277	39¾	50¼				1.2293	34½	55½
1.230-35	1.2344	43¼	46¾	1.2340	40	50	1.2318	37¼	52¾			
1.235-40	1.2388	43½	46½				1.2388	37½	52½	1.2371	34¾	55¼
1.240-45	1.2445	43¾	46¼	1.2404	40¼	49¾				1.2449	35	55
1.245-50	1.			1.2469	40½	49½	1.2458	37¾	52¼			
1.250-55	1.2502	44	46	1.2532	40¾	49¼	1.2528	38	52	1.2527	35¼	54¾
1.255-60	1.2559	44¼	45¾	1.2595	41	49	1.2598	38¼	51¾			
1.260-65	1.2615	44½	45½							1.2604	35½	54½
1.265-70	1.2671	44¾	45¼	1.2659	41¼	48¾	1.2668	38½	51½	1.2680	35¾	54¼
1.270-75	1.2727	45	45	1.2721	41½	48½	1.2737	38¾	51¼			
1.275-80	1.2783	45¼	44¾	1.2784	41¾	48¼				1.2758	36	54
1.280-85	1.2838	45½	44½	1.2847	42	48	1.2805	39	51	1.2833	36¼	53¾
1.285-90	1.2892	45¾	44¼				1.2875	39¼	50¾			
1.290-95	1.2947	46	44	1.2909	42¼	47¾	1.2943	39½	50½	1.2910	36½	53½
1.295-00				1.2976	42½	47½				1.2986	36¾	53¼
1.300-05	1.3001	46¼	43¾	1.3034	42¾	47¼	1.3012	39¾	50¼			
1.305-10	1.3055	46½	43½	1.3093	43	47	1.3080	40	50	1.3062	37	53
1.310-15	1.3110	46¾	43¼				1.3147	40¼	49¾	1.3138	37¼	52¾
1.315-20	1.3163	47	43	1.3167	43¼	46¾						
1.320-25	1.3217	47¼	42¾	1.3213	43½	46½	1.3216	40½	49½	1.3212	37½	52½
1.325-30	1.3270	47½	42½	1.3275	43¾	46¼	1.3283	40¾	49¼	1.3278	37¾	52¼
1.330-35	1.3322	47¾	42¼	1.3337	44	46						
1.335-40	1.3377	48	42	1.3396	44¼	45¾	1.3350	41	49	1.3362	38	52
1.340-45	1.3427	48¼	41¾				1.3417	41¼	48¾	1.3437	38¼	51¾
1.345-50	1.3480	48½	41½	1.3456	44½	45½	1.3483	41½	48½			

LEADS FROM 1.350 TO 1.500

Approximate Lead	Gear on Worm 28, 1st on Stud 56, 2nd on Stud 32, Gear on Screw 100	Angle to Set Spiral Head (Degrees)	Angle to Set Vertical Att. (Degrees)	Gear on Worm 24, 1st on Stud 72, 2nd on Stud 44, Gear on Screw 86	Angle to Set Spiral Head (Degrees)	Angle to Set Vertical Att. (Degrees)	Gear on Worm 24, 1st on Stud 64, 2nd on Stud 48, Gear on Screw 100	Angle to Set Spiral Head (Degrees)	Angle to Set Vertical Att. (Degrees)	Gear on Worm 24, 1st on Stud 40, 2nd on Stud 32, Gear on Screw 100	Angle to Set Spiral Head (Degrees)	Angle to Set Vertical Att. (Degrees)	Gear on Worm 28, 1st on Stud 64, 2nd on Stud 40, Gear on Screw 86	Angle to Set Spiral Head (Degrees)	Angle to Set Vertical Att. (Degrees)	Gear on Worm 24, 1st on Stud 72, 2nd on Stud 56, Gear on Screw 86	Angle to Set Spiral Head (Degrees)	Angle to Set Vertical Att. (Degrees)	Gear on Worm 24, 1st on Stud 64, 2nd on Stud 44, Gear on Screw 72	Angle to Set Spiral Head (Degrees)	Angle to Set Vertical Att. (Degrees)	Gear on Worm 28, 1st on Stud 64, 2nd on Stud 56, Gear on Screw 100	Angle to Set Spiral Head (Degrees)	Angle to Set Vertical Att. (Degrees)
1.350-55	1.3530	57¾	32¼	1.3530	52½	37½	1.3533	48¾	41¼	1.3512	44¾	45¼				1.3511	38½	51½				1.3522	33½	56½
1.355-60	1.3570	58	32	1.3576	52¾	37¼	1.3583	49	41	1.3576	45	45	1.3550	41¾	48¼	1.3585	38¾	51¼	1.3550	36¼	53¾			
1.360-65	1.3641	58½	31½	1.3620	53	37	1.3635	49¼	40¾	1.3635	45¼	44¾	1.3617	42	48				1.3630	36½	53½	1.3611	33¾	56¼
1.365-70	1.3676	58¾	31¼	1.3665	53¼	36¾	1.3686	49½	40½	1.3693	45½	44½	1.3682	42¼	47¾	1.3659	39	51						
1.370-75	1.3713	59	31	1.3710	53½	36½	1.3738	49¾	40¼				1.3748	42½	47½	1.3732	39¼	50¾	1.3710	36¾	53¼	1.3700	34	56
1.375-80	1.3784	59½	30½	1.3798	54	36	1.3788	50	40	1.3751	45¾	44¼							1.3790	37	53	1.3789	34¼	55¾
1.380-85	1.3819	59¾	30¼	1.3840	54¼	35¾	1.3838	50¼	39¾	1.3810	46	44	1.3815	42¾	47¼	1.3806	39½	50½						
1.385-90	1.3855	60	30	1.3883	54½	35½	1.3888	50½	39½	1.3868	46¼	43¾	1.3878	43	47	1.3880	39¾	50¼	1.3870	37¼	52¾	1.3877	34½	55½
1.390-95	1.3925	60½	29½	1.3927	54¾	35¼	1.3938	50¾	39¼	1.3925	46½	43½												
1.395-00	1.3992	61	29	1.3970	55	35	1.3987	51	39	1.3983	46¾	43¼	1.3956	43¼	46¾	1.3951	40	50	1.3950	37½	52½	1.3964	34¾	55¼
1.400-05	1.4025	61¼	28¾	1.4012	55¼	34¾	1.4036	51¼	38¾	1.4040	47	43	1.4005	43½	46½	1.4023	40¼	49¾	1.4029	37¾	52¼			
1.405-10	1.4059	61½	28½	1.4055	55½	34½	1.4085	51½	38½	1.4098	47¼	42¾	1.4071	43¾	46¼	1.4096	40½	49½				1.4052	35	55
1.410-15	1.4128	62	28	1.4139	56	34	1.4134	51¾	38¼				1.4135	44	46				1.4108	33	52	1.4140	35¼	54¾
1.415-20	1.4190	62½	27½	1.4180	56¼	33¾	1.4182	52	38	1.4155	47½	42½	1.4199	44¼	45¾	1.4168	40¾	49¼	1.4187	33¼	51¾			
1.420-25	1.4223	62¾	27¼	1.4220	56½	33½	1.4233	52¼	37¾	1.4212	47¾	42¼				1.4239	41	49				1.4227	35½	54½
1.425-30	1.4254	63	27	1.4267	56¾	33¼	1.4280	52½	37½	1.4267	48	42	1.4262	44½	45½				1.4265	38½	51½			
1.430-35	1.4318	63½	26½	1.4304	57	33	1.4328	52¾	37¼	1.4322	48¼	41¾	1.4326	44¾	45¼	1.4310	41¼	48¾	1.4342	38¾	51¼	1.4313	35¾	54¼
1.435-40	1.4380	64	26	1.4385	57½	32½	1.4375	53	37	1.4379	48½	41½	1.4389	45	45	1.4381	41½	48½						
1.440-45	1.4441	64½	25½	1.4423	57¾	32¼	1.4422	53¼	36¾	1.4435	48¾	41¼							1.4420	39	51	1.4400	36	54
1.445-50	1.4499	65	25	1.4462	58	32	1.4469	53½	36½	1.4490	49	41	1.4451	45¼	44¾	1.4452	41¾	48¼	1.4499	39¼	50¾	1.4487	36¼	53¾
1.450-55	1.4529	65¼	24¾	1.4540	58½	31½	1.4515	53¾	36¼	1.4544	49¼	40¾	1.4513	45½	44½	1.4523	42	48						
1.455-60	1.4558	65½	24½	1.4579	58¾	31¼	1.4561	54	36	1.4599	49½	40½	1.4576	45¾	44¼	1.4593	42¼	47¾	1.4576	39½	50½	1.4572	36½	53½
1.460-65	1.4617	66	24	1.4619	59	31	1.4607	54¼	35¾				1.4637	46	44									
1.465-70	1.4672	66½	23½	1.4693	59½	30½	1.4652	54½	35½	1.4652	49¾	40¼	1.4699	46¼	43¾	1.4673	42½	47½	1.4652	39¾	50¼	1.4658	36¾	53¼
1.470-75	1.4726	67	23	1.4731	59¾	30¼	1.4745	55	35	1.4707	50	40				1.4735	42¾	47¼	1.4730	40	50	1.4743	37	53
1.475-80	1.4780	67½	22½	1.4769	60	30	1.4789	55¼	34¾	1.4760	50¼	39¾	1.4759	46½	43½									
1.480-85	1.4835	68	22	1.4844	60½	29½	1.4833	55½	34½	1.4813	50½	39½	1.4821	46¾	43¼	1.4802	43	47	1.4805	40¼	49¾	1.4829	37¼	52¾
1.485-90	1.4885	68½	21½	1.4879	60¾	29¼	1.4877	55¾	34¼	1.4867	50¾	39¼	1.4881	47	43	1.4886	43¼	46¾	1.4882	40½	49½			
1.490-95	1.4938	69	21	1.4916	61	29	1.4921	56	34	1.4919	51	39	1.4942	47¼	42¾							1.4914	37½	52½
1.495-00	1.4988	69½	20½	1.4987	61½	28½	1.4964	56¼	33¾	1.4972	51¼	38¾				1.4958	43½	46½	1.4958	40¾	49¼	1.4999	37¾	52¼

LEADS FROM 1.500 TO 1.650

Approximate Lead	Gear on Worm 24, 1st on Stud 72, 2nd on Stud 44, Gear on Screw 86	Angle to Set Spiral Head (Degrees)	Angle to Set Vertical Att. (Degrees)	Gear on Worm 24, 1st on Stud 64, 2nd on Stud 48, Gear on Screw 100	Angle to Set Spiral Head (Degrees)	Angle to Set Vertical Att. (Degrees)	Gear on Worm 24, 1st on Stud 40, 2nd on Stud 32, Gear on Screw 100	Angle to Set Spiral Head (Degrees)	Angle to Set Vertical Att. (Degrees)	Gear on Worm 28, 1st on Stud 64, 2nd on Stud 40, Gear on Screw 86	Angle to Set Spiral Head (Degrees)	Angle to Set Vertical Att. (Degrees)
1.500-05	1.5025	61¾	28¼	1.5008	56½	33½	1.5025	51½	38½	1.5002	47½	42½
1.505-10	1.5069	62	28	1.5096	57	33	1.5076	51¾	38¼	1.5061	47¾	42¼
1.510-15	1.5127	62½	27½	1.5137	57¼	32¾	1.5128	52	38	1.5124	48	42
1.515-20	1.5196	63	27	1.5180	57½	32½	1.5181	52¼	37¾	1.5181	48¼	41¾
1.520-25	1.5230	63¼	26¾	1.5221	57¾	32¼	1.5232	52½	37½	1.5240	48½	41½
1.525-30	1.5261	63½	26½	1.5262	58	32	1.5283	52¾	37¼			
1.530-35	1.5329	64	26	1.5345	58½	31½	1.5334	53	37	1.5300	48¾	41¼
1.535-40	1.5393	64½	25½	1.5386	58¾	31¼	1.5383	53¼	36¾	1.5357	49	41
1.540-45	1.5425	64¾	25¼	1.5428	59	31	1.5433	53½	36½	1.5416	49¼	40¾
1.545-50	1.5456	65	25	1.5468	59¼	30¾	1.5488	53¾	36¼	1.5472	49½	40½
1.550-55	1.5519	65½	24½	1.5506	59½	30½	1.5532	54	36	1.5531	49¾	40¼
1.555-60	1.5580	66	24	1.5588	60	30	1.5580	54¼	35¾	1.5588	50	40
1.560-65	1.5640	66½	23½	1.5626	60¼	29¾	1.5630	54½	35½	1.5645	50¼	39¾
1.565-70	1.5698	67	23	1.5665	60½	29½	1.5678	54¾	35¼			
1.570-75	1.5727	67¼	22¾	1.5741	61	29	1.5726	55	35	1.5700	50½	39½
1.575-80	1.5755	67½	22½	1.5779	61¼	28¾	1.5774	55¼	34¾	1.5758	50¾	39¼
1.580-85	1.5810	68	22	1.5816	61½	28½	1.5822	55½	34½	1.5812	51	39
1.585-90	1.5877	68½	21½	1.5893	62	28	1.5869	55¾	34¼	1.5869	51¼	38¾
1.590-95	1.5920	69	21	1.5929	62¼	27¾	1.5916	56	34	1.5925	51½	38½
1.595-00	1.5973	69½	20½	1.5964	62½	27½	1.5962	56¼	33¾	1.5979	51¾	38¼
1.600-05	1.6025	70	20	1.6037	63	27	1.6009	56½	33½	1.6024	52	38
1.605-10	1.6075	70½	19½	1.6074	63¼	26¾	1.6060	56¾	33¼	1.6091	52¼	37¾
1.610-15	1.6125	71	19	1.6108	63½	26½	1.6102	57	33	1.6145	52½	37½
1.615-20	1.6172	71½	18½	1.6178	64	26	1.6192	57½	32½	1.6199	52¾	37¼
1.620-25	1.6220	72	18	1.6246	64½	25½	1.6237	57¾	32¼			
1.625-30	1.6264	72½	17½	1.6279	64¾	25¼	1.6280	58	32	1.6252	53	37
1.630-35	1.6309	73	17	1.6311	65	25	1.6325	58¼	31¾	1.6305	53¼	36¾
1.635-40	1.6393	74	16	1.6379	65½	24½	1.6369	58½	31½	1.6358	53½	36½
1.640-45	1.6433	74½	15½	1.6442	66	24	1.6411	58¾	31¼	1.6410	53¾	36¼
1.645-50	1.6471	75	15	1.6474	66¼	23¾	1.6457	59	31	1.6462	54	36

Approximate Lead	Gear on Worm 24, 1st on Stud 72, 2nd on Stud 56, Gear on Screw 86	Angle to Set Spiral Head (Degrees)	Angle to Set Vertical Att. (Degrees)	Gear on Worm 24, 1st on Stud 64, 2nd on Stud 44, Gear on Screw 72	Angle to Set Spiral Head (Degrees)	Angle to Set Vertical Att. (Degrees)	Gear on Worm 28, 1st on Stud 64, 2nd on Stud 56, Gear on Screw 100	Angle to Set Spiral Head (Degrees)	Angle to Set Vertical Att. (Degrees)	Gear on Worm 40, 1st on Stud 86, 2nd on Stud 56, Gear on Screw 100	Angle to Set Spiral Head (Degrees)	Angle to Set Vertical Att. (Degrees)
1.500-05	1.5009	43¾	46¼	1.5033	41	49				1.5034	35¼	54¾
1.505-10	1.5077	44	46				1.5083	38	52	1.5127	35½	54½
1.510-15	1.5145	44¼	45¾	1.5109	41¼	48¾						
1.515-20				1.5183	41½	48½	1.5167	38¼	51¾			
1.520-25	1.5212	44½	45½							1.5219	35¾	54¼
1.525-30	1.5280	44¾	45¼	1.5259	41¾	48¼	1.5251	38½	51½			
1.530-35	1.5348	45	45	1.5333	42	48	1.5334	38¾	51¼	1.5311	36	54
1.535-40												
1.540-45	1.5414	45¼	44¾	1.5408	42¼	47¾	1.5418	39	51	1.5402	36¼	53¾
1.545-50	1.5480	45½	44½	1.5481	42½	47½				1.5494	36½	53½
1.550-55	1.5546	45¾	44¼				1.5500	39¼	50¾			
1.555-60				1.5558	42¾	47¼	1.5583	39½	50½	1.5585	36¾	53¼
1.560-65	1.5612	46	44	1.5628	43	47						
1.565-70	1.5678	46¼	43¾				1.5666	39¾	50¼	1.5677	37	53
1.570-75	1.5742	46½	43½	1.5716	43¼	46¾	1.5748	40	50			
1.575-80				1.5771	43½	46½				1.5767	37¼	52¾
1.580-85	1.5808	46¾	43¼	1.5845	43¾	46¼	1.5829	40¼	49¾			
1.585-90	1.5873	47	43							1.5857	37½	52½
1.590-95	1.5938	47¼	42¾	1.5918	44	46	1.5911	40½	49½	1.5948	37¾	52¼
1.595-00				1.5990	44¼	45¾	1.5991	40¾	49¼			
1.600-05	1.6002	47½	42½							1.6038	38	52
1.605-10	1.6065	47¾	42¼	1.6061	44½	45½	1.6073	41	49			
1.610-15	1.6131	48	42	1.6131	44¾	45¼				1.6127	38¼	51¾
1.615-20	1.6192	48¼	41¾				1.6153	41¼	48¾			
1.620-25				1.6204	45	45	1.6233	41½	48½	1.6216	38½	51½
1.625-30	1.6254	48½	41½	1.6274	45¼	44¾						
1.630-35	1.6319	48¾	41¼	1.6345	45½	44½	1.6313	41¾	48¼	1.6304	38¾	51¼
1.635-40	1.6380	49	41				1.6392	42	48	1.6392	39	51
1.640-45	1.6442	49¼	40¾	1.6413	45¾	44¼						
1.645-50				1.6483	46	44	1.6472	42¼	47¾	1.6481	39¼	50¾

LEADS FROM 1.650 TO 1.800

Approximate Lead	Gear on Worm 24, 1st on Stud 40, 2nd on Stud 32, Gear on Screw 100	Angle to Set Spiral Head (Degrees)	Angle to Set Vertical Att. (Degrees)	Gear on Worm 28, 1st on Stud 64, 2nd on Stud 40, Gear on Screw 86	Angle to Set Spiral Head (Degrees)	Angle to Set Vertical Att. (Degrees)	Gear on Worm 24, 1st on Stud 72, 2nd on Stud 56, Gear on Screw 86	Angle to Set Spiral Head (Degrees)	Angle to Set Vertical Att. (Degrees)	Gear on Worm 24, 1st on Stud 64, 2nd on Stud 44, Gear on Screw 72	Angle to Set Spiral Head (Degrees)	Angle to Set Vertical Att. (Degrees)	Gear on Worm 28, 1st on Stud 64, 2nd on Stud 56, Gear on Screw 100	Angle to Set Spiral Head (Degrees)	Angle to Set Vertical Att. (Degrees)	Gear on Worm 40, 1st on Stud 86, 2nd on Stud 56, Gear on Screw 100	Angle to Set Spiral Head (Degrees)	Angle to Set Vertical Att. (Degrees)	Gear on Worm 40, 1st on Stud 56, 2nd on Stud 28, Gear on Screw 72	Angle to Set Spiral Head (Degrees)	Angle to Set Vertical Att. (Degrees)	Gear on Worm 24, 1st on Stud 56, 2nd on Stud 44, Gear on Screw 64	Angle to Set Spiral Head (Degrees)	Angle to Set Vertical Att. (Degrees)
1.650-55	1.6541	59½	30½	1.6514	54¼	35¾	1.6503	49½	40½										1.6523	36½	53½			
1.655-60	1.6582	59¾	30¼	1.6566	54½	35½	1.6565	49¾	40¼	1.6553	46¼	43¾	1.6551	42½	47½	1.6570	39½	50½				1.6581	34¼	55¾
1.660-65	1.6625	60	30	1.6617	54¾	35¼	1.6627	50	40	1.6620	46½	43½	1.6632	42¾	47¼				1.6620	36¾	53¼			
1.665-70	1.6667	60¼	29¾	1.6669	55	35	1.6687	50¼	39¾	1.6690	46¾	43¼				1.6657	39¾	50¼				1.6688	34½	55½
1.670-75	1.6710	60½	29½	1.6720	55¼	34¾	1.6748	50½	39½				1.6708	43	47	1.6743	40	50	1.6718	37	53			
1.675-80	1.6791	61	29	1.6771	55½	34½				1.6759	47	43										1.6793	34¾	55¼
1.680-85	1.6831	61¼	28¾	1.6820	55¾	34¼	1.6808	50¾	39¼	1.6827	47¼	42¾	1.6803	43¼	46¾	1.6830	40¼	49¾	1.6814	37¼	52¾			
1.685-90	1.6870	61½	28½	1.6870	56	34	1.6866	51	39	1.6895	47½	42½	1.6861	43½	46½							1.6899	35	55
1.690-95	1.6912	61¾	28¼	1.6919	56¼	33¾	1.6927	51¼	38¾				1.6940	43¾	46¼	1.6918	40½	49½	1.6911	37½	52½			
1.695-00	1.6952	62	28	1.6968	56½	33½	1.6985	51½	38½	1.6960	47¾	42¼												
1.700-05	1.7029	62½	27½	1.7023	56¾	33¼	1.7044	51¾	38¼	1.7031	48	42	1.7019	44	46	1.7003	40¾	49¼	1.7006	37¾	52¼	1.7004	35¼	54¾
1.705-10	1.7068	62¾	27¼	1.7068	57	33				1.7095	48¼	41¾	1.7094	44¼	45¾	1.7090	41	49						
1.710-15	1.7106	63	27	1.7112	57¼	32¾	1.7103	52	38										1.7102	38	52	1.7109	35½	54½
1.715-20	1.7181	63½	26½	1.7162	57½	32½	1.7164	52¼	37¾	1.7160	48½	41½	1.7171	44½	45½	1.7175	41¼	48¾	1.7198	38¼	51¾			
1.720-25	1.7220	63¾	26¼	1.7209	57¾	32¼	1.7220	52½	37½	1.7230	48¾	41¼	1.7248	44¾	45¼							1.7211	35¾	54¼
1.725-30	1.7257	64	26	1.7255	58	32	1.7278	52¾	37¼	1.7294	49	41				1.7260	41½	48½	1.7292	38½	51½			
1.730-35	1.7329	64½	25½	1.7349	58½	31½	1.7334	53	37				1.7324	45	45	1.7345	41¾	48¼				1.7317	36	54
1.735-40	1.7399	65	25	1.7395	58¾	31¼	1.7390	53¼	36¾	1.7360	49¼	40¾	1.7399	45¼	44¾				1.7388	38¾	51¼			
1.740-45	1.7434	65¼	24¾	1.7442	59	31	1.7448	53½	36½	1.7424	49½	40½				1.7430	42	48				1.7420	36¼	53¾
1.745-50	1.7470	65½	24½	1.7488	59¼	30¾				1.7490	49¾	40¼	1.7474	45½	44½				1.7481	39	51			
1.750-55	1.7539	66	24	1.7531	59½	30½	1.7503	53¾	36¼				1.7549	45¾	44¼	1.7515	42¼	47¾				1.7524	36½	53½
1.755-60	1.7571	66¼	23¾	1.7578	59¾	30¼	1.7560	54	36	1.7553	50	40				1.7599	42½	47½	1.7575	39¼	50¾			
1.760-65	1.7608	66½	23½	1.7621	60	30	1.7614	54¼	35¾	1.7618	50¼	39¾	1.7623	46	44							1.7627	36¾	53¼
1.765-70	1.7671	67	23	1.7667	60¼	29¾	1.7670	54½	35½	1.7681	50½	39½	1.7697	46¼	43¾				1.7670	39½	50½			
1.770-75	1.7737	67½	22½	1.7711	60½	29½	1.7725	54¾	35¼	1.7745	50¾	39¼				1.7684	42¾	47½				1.7730	37	53
1.775-80	1.7771	67¾	22¼	1.7799	61	29	1.7779	55	35				1.7770	46½	43½	1.7765	43	47	1.7763	39¾	50¼			
1.780-85	1.7800	68	22	1.7840	61¼	28¾	1.7833	55¼	34¾	1.7808	51	39	1.7849	46¾	43¼							1.7833	37¼	52¾
1.785-90	1.7862	68½	21½	1.7881	61½	28½	1.7888	55½	34½	1.7871	51¼	38¾				1.7866	43¼	46¾	1.7857	40	50			
1.790-95	1.7922	69	21	1.7925	61¾	28¼	1.7940	55¾	34¼	1.7933	51½	38½	1.7918	47	43				1.7949	40¼	49¾	1.7935	37½	52½
1.795-00	1.7981	69½	20½	1.7969	62	28	1.7993	56	34	1.7996	51¾	38¼	1.7990	47¼	42¾	1.7962	43½	46½						

Approximate Lead	Gear on Worm 28, 1st on Stud 64, 2nd on Stud 40, Gear on Screw 86	Angle to Set Spiral Head (Degrees)	Angle to Set Vertical Att. (Degrees)	Gear on Worm 24, 1st on Stud 72, 2nd on Stud 56, Gear on Screw 86	Angle to Set Spiral Head (Degrees)	Angle to Set Vertical Att. (Degrees)	Gear on Worm 24, 1st on Stud 64, 2nd on Stud 44, Gear on Screw 72	Angle to Set Spiral Head (Degrees)	Angle to Set Vertical Att. (Degrees)	Gear on Worm 28, 1st on Stud 64, 2nd on Stud 56, Gear on Screw 100	Angle to Set Spiral Head (Degrees)	Angle to Set Vertical Att. (Degrees)	Gear on Worm 40, 1st on Stud 86, 2nd on Stud 56, Gear on Screw 100	Angle to Set Spiral Head (Degrees)	Angle to Set Vertical Att. (Degrees)	Gear on Worm 40, 1st on Stud 56, 2nd on Stud 28, Gear on Screw 72	Angle to Set Spiral Head (Degrees)	Angle to Set Vertical Att. (Degrees)	Gear on Worm 24, 1st on Stud 56, 2nd on Stud 44, Gear on Screw 64	Angle to Set Spiral Head (Degrees)	Angle to Set Vertical Att. (Degrees)	Gear on Worm 24, 1st on Stud 86, 2nd on Stud 72, Gear on Screw 64	Angle to Set Spiral Head (Degrees)	Angle to Set Vertical Att. (Degrees)
1.800-05	1.8049	62½	27½	1.8046	56¼	33¾							1.8012	43¾	46¼	1.8041	40½	49½	1.8038	37¾	52¼	1.8008	35	55
1.805-10	1.8090	62¾	27¼	1.8099	56½	33½	1.8058	52	38	1.8063	47½	42½	1.8095	44	46									
1.810-15	1.8130	63	27				1.8121	52¼	37¾	1.8133	47¾	42¼				1.8132	40¾	49¼	1.8140	38	52	1.8120	35¼	54¾
1.815-20	1.8172	63¼	26¾	1.8157	56¾	33¼	1.8181	52½	37½				1.8176	44¼	45¾									
1.820-25	1.8211	63½	26½	1.8204	57	33	1.8242	52¾	37¼	1.8209	48	42				1.8224	41	49	1.8241	38¼	51¾	1.8231	35½	54½
1.825-30	1.8290	64	26	1.8254	57¼	32¾				1.8278	48¼	41¾	1.8258	44½	45½									
1.830-35	1.8329	64¼	25¾	1.8307	57½	32½	1.8301	53	37	1.8348	48½	41½	1.8339	44¾	45¼	1.8316	41¼	48¾	1.8341	38½	51½	1.8342	35¾	54¼
1.835-40	1.8368	64½	25½	1.8355	57¾	32¼	1.8361	53¼	36¾															
1.840-45	1.8441	65	25	1.8405	58	32	1.8421	53½	36½	1.8420	48¾	41¼	1.8420	45	45	1.8407	41½	48½	1.8441	38¾	51¼			
1.845-50	1.8479	65¼	24¾	1.8455	58¼	31¾	1.8480	53¾	36¼	1.8490	49	41				1.8498	41¾	48¼				1.8453	36	54
1.850-55	1.8517	65½	24½	1.8505	58½	31½	1.8540	54	36				1.8500	45¼	44¾				1.8541	39	51			
1.855-60	1.8590	66	24	1.8555	58¾	31¼	1.8598	54¼	35¾	1.8560	49¼	40¾	1.8580	45½	44½	1.8588	42	48				1.8563	36¼	53¾
1.860-65	1.8625	66¼	23¾	1.8605	59	31				1.8629	49½	40½							1.8642	39¼	50¾			
1.865-70	1.8661	66½	23½	1.8699	59½	30½	1.8655	54½	35½	1.8690	49¾	40¼	1.8659	45¾	44¼	1.8678	42¼	47¾				1.8674	36½	53½
1.870-75	1.8730	67	23	1.8748	59¾	30¼	1.8713	54¾	35¼				1.8737	46	44				1.8742	39½	50½			
1.875-80	1.8765	67¼	22¾	1.8797	60	30	1.8770	55	35	1.8768	50	40				1.8768	42½	47½				1.8784	36¾	53¼
1.880-85	1.8800	67½	22½	1.8845	60¼	29¾	1.8829	55¼	34¾	1.8836	50¼	39¾	1.8817	46¼	43¾				1.8841	39¾	50¼			
1.885-90	1.8867	68	22	1.8891	60½	29½	1.8887	55½	34½				1.8894	46½	43½	1.8859	42¾	47¼				1.8894	37	53
1.890-95	1.8932	68½	21½	1.8936	60¾	29¼	1.8941	55¾	34¼	1.8904	50½	39½				1.8945	43	47	1.8939	40	50			
1.895-00	1.8997	69	21	1.8982	61	29	1.8998	56	34	1.8972	50¾	39¼	1.8973	46¾	43¼									
1.900-05	1.9029	69¼	20¾	1.9028	61¼	28¾				1.9038	51	39							1.9035	40¼	49¾	1.9004	37¼	52¾
1.905-10	1.9060	69½	20½	1.9071	61½	28½	1.9053	56¼	33¾				1.9052	47	43	1.9052	43¼	46¾						
1.910-15	1.9122	70	20	1.9120	61¾	28¼	1.9109	56½	33½	1.9106	51¼	38¾	1.9129	47¼	42¾	1.9123	43½	46½	1.9135	40½	49½	1.9113	37½	52½
1.915-20	1.9181	70½	19½	1.9166	62	28	1.9170	56¾	33¼	1.9172	51½	38½												
1.920-25	1.9240	71	19	1.9210	62¼	27¾	1.9219	57	33	1.9239	51¾	38¼	1.9205	47½	42½	1.9210	43¾	46¼	1.9232	40¾	49¼	1.9221	37¾	52¼
1.925-30	1.9298	71½	18½	1.9251	62½	27½	1.9271	57¼	32¾				1.9280	47¾	42¼	1.9297	44	46						
1.930-35	1.9326	71¾	18¼	1.9338	63	27	1.9328	57½	32½	1.9305	52	38							1.9329	41	49	1.9329	38	52
1.935-40	1.9355	72	18	1.9382	63¼	26¾	1.9380	57¾	32¼	1.9373	52¼	37¾	1.9360	48	42	1.9383	44¼	45¾						
1.940-45	1.9407	72½	17½	1.9424	63½	26½	1.9431	58	32	1.9438	52½	37½	1.9433	48¼	41¾				1.9426	41¼	48¾	1.9437	38¼	51¾
1.945-50	1.9460	73	17	1.9467	63¾	26¼	1.9485	58¼	31¾							1.9470	44½	45½						

LEADS FROM 1.950 TO 2.100

Approximate Lead	Gear on Worm 24, 1st on Stud 72, 2nd on Stud 56, Gear on Screw 86	Angle to Set Spiral Head (Degrees)	Angle to Set Vertical Att. (Degrees)	Gear on Worm 24, 1st on Stud 64, 2nd on Stud 44, Gear on Screw 72	Angle to Set Spiral Head (Degrees)	Angle to Set Vertical Att. (Degrees)	Gear on Worm 28, 1st on Stud 64, 2nd on Stud 56, Gear on Screw 100	Angle to Set Spiral Head (Degrees)	Angle to Set Vertical Att. (Degrees)	Gear on Worm 40, 1st on Stud 86, 2nd on Stud 56, Gear on Screw 100	Angle to Set Spiral Head (Degrees)	Angle to Set Vertical Att. (Degrees)	Gear on Worm 40, 1st on Stud 56, 2nd on Stud 28, Gear on Screw 72	Angle to Set Spiral Head (Degrees)	Angle to Set Vertical Att. (Degrees)	Gear on Worm 24, 1st on Stud 56, 2nd on Stud 44, Gear on Screw 64	Angle to Set Spiral Head (Degrees)	Angle to Set Vertical Att. (Degrees)	Gear on Worm 24, 1st on Stud 86, 2nd on Stud 72, Gear on Screw 64	Angle to Set Spiral Head (Degrees)	Angle to Set Vertical Att. (Degrees)	Gear on Worm 28, 1st on Stud 56, 2nd on Stud 48, Gear on Screw 72	Angle to Set Spiral Head (Degrees)	Angle to Set Vertical Att. (Degrees)
1.950-55	1.9509	64	26	1.9538	58½	31½	1.9502	52¾	37¼	1.9511	48½	41½				1.9523	41½	48½	1.9544	38½	51½			
1.955-60	1.9590	64½	25½	1.9590	58¾	31¼	1.9567	53	37	1.9585	48¾	41¼	1.9557	44¾	45¼							1.9592	36	54
1.960-65	1.9630	64¾	25¼	1.9642	59	31	1.9630	53¼	36¾				1.9642	45	45	1.9619	41¾	48¼						
1.965-70	1.9670	65	25	1.9692	59¼	30¾	1.9694	53½	36½	1.9660	49	41							1.9650	38¾	51¼			
1.970-75	1.9710	65¼	24¾	1.9744	59½	30½				1.9734	49¼	40¾	1.9728	45¼	44¾	1.9714	42	48				1.9709	36¼	53¾
1.975-80	1.9750	65½	24½	1.9794	59¾	30¼	1.9758	53¾	36¼										1.9757	39	51			
1.980-85	1.9829	66	24	1.9845	60	30	1.9820	54	36	1.9808	49½	40½	1.9815	45½	44½	1.9810	42¼	47¾				1.9827	36½	53½
1.985-90	1.9865	66¼	23¾	1.9896	60¼	29¾	1.9882	54¼	35¾	1.9881	49¾	40¼	1.9899	45¾	44¼				1.9864	39¼	50¾			
1.990-95	1.9906	66½	23½	1.9945	60½	29½	1.9945	54½	35½							1.9904	42½	47½				1.9943	36¾	53¼
1.995-00	1.9979	67	23	1.9995	60¾	29¼				1.9955	50	40	1.9983	46	44	2.000	42¾	47¼	1.9970	39½	50½			
2.000-05	2.0016	67¼	22¾	2.0041	61	29	2.0007	54¾	35¼	2.0028	50¼	39¾												
2.005-10	2.0051	67½	22½	2.0090	61¼	28¾	2.0069	55	35				2.0066	46¼	43¾	2.0094	43	47	2.0075	39¾	50¼	2.0060	37	53
2.010-15	2.0123	68	22	2.0138	61½	28½	2.0130	55¼	34¾	2.0100	50½	39½	2.0149	46½	43½									
2.015-20	2.0193	68½	21½	2.0187	61¾	28¼	2.0191	55½	34½	2.0172	50¾	39¼				2.0187	43¼	46¾	2.0181	40	50	2.0177	37¼	52¾
2.020-25	2.0229	68¾	21¼	2.0236	62	28				2.0247	51	39	2.0233	46¾	43¼									
2.025-30	2.0264	69	21	2.0280	62¼	27¾	2.0250	55¾	34¼							2.0282	43½	46½	2.0286	40¼	49¾	2.0291	37½	52½
2.030-35	2.0329	69½	20½	2.0325	62½	27½	2.0311	56	34	2.0315	51¼	38¾	2.0319	47	43									
2.035-40	2.0395	70	20	2.0371	62¾	27¼	2.0370	56¼	33¾	2.0386	51½	38½				2.0373	43¾	46¼	2.0390	40½	49½			
2.040-45	2.0428	70¼	19¾	2.0418	63	27	2.0429	56½	33½				2.0400	47¼	42¾							2.0407	37¾	52¼
2.045-50	2.0459	70½	19½	2.0465	63¼	26¾	2.0495	56¾	33¼	2.0456	51¾	38¼	2.0481	47½	42½	2.0467	44	46	2.0494	40¾	49¼			
2.050-55	2.0521	71	19	2.0507	63½	26½	2.0549	57	33	2.0525	52	38										2.0521	38	52
2.055-60	2.0583	71½	18½	2.0598	64	26							2.0561	47¾	42¼	2.0558	44¼	45¾						
2.060-65	2.0644	72	18	2.0640	64¼	25¾	2.0602	57¼	32¾	2.0600	52¼	37¾	2.0643	48	42	2.0651	44½	45½	2.0597	41	49	2.0635	38¼	51¾
2.065-70	2.0672	72¼	17¾	2.0684	64½	25½	2.0663	57½	32½	2.0667	52½	37½												
2.070-75	2.0700	72½	17½	2.0726	64¾	25¼	2.0719	57¾	32¼	2.0736	52¾	37¼	2.0723	48¼	41¾	2.0742	44¾	45¼	2.0700	41¼	48¾	2.0749	38½	51½
2.075-80	2.0757	73	17	2.0768	65	25	2.0774	58	32															
2.080-85	2.0811	73½	16½	2.0810	65¼	24¾	2.0831	58¼	31¾	2.0805	53	37	2.0808	48½	41½	2.0834	45	45	2.0803	41½	48½			
2.085-90	2.0865	74	16	2.0851	65½	24½	2.0889	58½	31½	2.0872	53¼	36¾	2.0886	48¾	41¼							2.0863	38¾	51¼
2.090-95	2.0915	74½	15½	2.0935	66	24										2.0924	45¼	44¾	2.0906	41¾	48¼			
2.095-00	2.0965	75	15	2.0975	66¼	23¾	2.0942	58¾	31¼	2.0940	53½	36½	2.0965	49	41							2.0976	39	51

LEADS FROM 2.100 TO 2.250

Approximate Lead	Gear on Worm 24, 1st on Stud 64, 2nd on Stud 44, Gear on Screw 72	Angle to Set Spiral Head (Degrees)	Angle to Set Vertical Att. (Degrees)	Gear on Worm 28, 1st on Stud 64, 2nd on Stud 56, Gear on Screw 100	Angle to Set Spiral Head (Degrees)	Angle to Set Vertical Att. (Degrees)	Gear on Worm 40, 1st on Stud 86, 2nd on Stud 56, Gear on Screw 100	Angle to Set Spiral Head (Degrees)	Angle to Set Vertical Att. (Degrees)	Gear on Worm 40, 1st on Stud 56, 2nd on Stud 28, Gear on Screw 72	Angle to Set Spiral Head (Degrees)	Angle to Set Vertical Att. (Degrees)
2.100-05	2.1017	66½	23½	2.1000	59	31	2.1007	53¾	36¼	2.1045	49¼	40¾
2.105-10	2.1093	67	23	2.1054	59¼	30¾	2.1075	54	36			
2.110-15	2.1131	67¼	22¾	2.1108	59½	30½	2.1140	54¼	35¾	2.1122	49½	40½
2.115-20	2.1170	67½	22½	2.1161	59¾	30¼						
2.120-25	2.1246	68	22	2.1216	60	30	2.1207	54½	35½	2.1201	49¾	40¼
2.125-30	2.1284	68¼	21¾	2.1270	60¼	29¾	2.1273	54¾	35¼	2.1280	50	40
2.130-35	2.1320	68½	21½	2.1324	60½	29½	2.1339	55	35			
2.135-40	2.1392	69	21	2.1375	60¾	29¼				2.1358	50¼	39¾
2.140-45	2.1430	69¼	20¾	2.1429	61	29	2.1403	55¼	34¾	2.1435	50½	39½
2.145-50	2.1463	69½	20½	2.1479	61¼	28¾	2.1469	55½	34½			
2.150-55	2.1533	70	20	2.1529	61½	28½	2.1531	55¾	34¼	2.1511	50¾	39¼
2.155-60	2.1567	70¼	19¾	2.1581	61¾	28¼	2.1595	56	34	2.1587	51	39
2.160-65	2.1600	70½	19½	2.1632	62	28						
2.165-70	2.1668	71	19	2.1682	62¼	27¾	2.1658	56¼	33¾	2.1664	51¼	38¾
2.170-75	2.1730	71½	18½	2.1730	62½	27½	2.1721	56½	33½	2.1740	51½	38½
2.175-80	2.1795	72	18	2.1780	62¾	27¼	2.1791	56¾	33¼			
2.180-85	2.1825	72¼	17¾	2.1829	63	27	2.1849	57	33	2.1814	51¾	38¼
2.185-90	2.1854	72½	17½	2.1879	63¼	26¾				2.1890	52	38
2.190-95	2.1914	73	17	2.1925	63½	26½	2.1907	57¼	32¾			
2.195-00	2.1971	73½	16½	2.1974	63¾	26¼	2.1970	57½	32½	2.1968	52¼	37¾
2.200-05	2.2029	74	16	2.2020	64	26	2.2030	57¾	32¼	2.204	52½	37½
2.205-10	2.2084	74½	15½	2.2068	64¼	25¾	2.2089	58	32			
2.210-15	2.2136	75	15	2.2112	64½	25½				2.2112	52¾	37¼
2.215-20	2.2187	75½	14½	2.2159	64¾	25¼	2.2150	58¼	31¾	2.2187	53	37
2.220-25	2.2237	76	14	2.2202	65	25	2.2209	58½	31½			
2.225-30	2.2283	76½	13½	2.2292	65½	24½	2.2269	58¾	31¼	2.2258	53¼	36¾
2.230-35	2.2328	77	13	2.2339	65¾	24¼	2.2329	59	31	2.2331	53½	36½
2.235-40	2.2372	77½	12½	2.2381	66	24	2.2385	59¼	30¾			
2.240-45	2.2415	78	12	2.2424	66¼	23¾	2.2443	59½	30½	2.2402	53¾	36¼
2.245-50	2.2457	78½	11½	2.2479	66½	23½				2.2474	54	36

Approximate Lead	Gear on Worm 24, 1st on Stud 56, 2nd on Stud 44, Gear on Screw 64	Angle to Set Spiral Head (Degrees)	Angle to Set Vertical Att. (Degrees)	Gear on Worm 24, 1st on Stud 86, 2nd on Stud 72, Gear on Screw 64	Angle to Set Spiral Head (Degrees)	Angle to Set Vertical Att. (Degrees)	Gear on Worm 28, 1st on Stud 56, 2nd on Stud 48, Gear on Screw 72	Angle to Set Spiral Head (Degrees)	Angle to Set Vertical Att. (Degrees)	Gear on Worm 48, 1st on Stud 44, 2nd on Stud 28, Gear on Screw 86	Angle to Set Spiral Head (Degrees)	Angle to Set Vertical Att. (Degrees)
2.100-05	2.1015	45½	44½	2.1008	42	48				2.1003	36¼	53¾
2.105-10							2.1090	39¼	50¾			
2.110-15	2.1105	45¾	44¼	2.1109	42¼	46¾	2.1102	39½	50½	2.1127	36½	53½
2.115-20	2.1193	46	44									
2.120-25				2.1210	42½	46½						
2.125-30	2.1283	46¼	43¾							2.1251	36¾	53¼
2.130-35				2.1311	42¾	46¼	2.1315	39¾	50¼			
2.135-40	2.1372	46½	43½							2.1376	37	53
2.140-45				2.1412	43	46	2.1426	40	50			
2.145-50	2.1461	46¾	43¼									
2.150-55	2.1549	47	43	2.1512	43¼	45¾	2.1537	40¼	49¾	2.1500	37¼	52¾
2.155-60												
2.160-65	2.1636	47¼	42¾	2.1611	43½	45½	2.1648	40½	49½	2.1622	37½	52½
2.165-70												
2.170-75	2.1724	47½	42½	2.1710	43¾	46¼				2.1745	37¾	52¼
2.175-80							2.1758	40¾	49¼			
2.180-85	2.1808	47¾	42¼	2.1809	44	46						
2.185-90	2.1896	48	42				2.1868	41	49	2.1861	38	52
2.190-95				2.1907	44¼	45¾						
2.195-00	2.1980	48¼	41¾				2.1978	41¼	48¾	2.1989	38¼	51¾
2.200-05				2.2005	44½	45½						
2.205-10	2.2066	48½	41½				2.2087	41½	48½			
2.210-15	2.2152	48¾	41¼	2.2103	44¾	45¼				2.2111	38½	51½
2.215-20							2.2195	41¾	48¼			
2.220-25	2.2236	49	41	2.2200	45	45				2.2232	38¾	51¼
2.225-30												
2.230-35	2.2320	49¼	40¾	2.2297	45¼	44¾	2.2303	42	48			
2.235-40				2.2393	45½	44½				2.2351	39	51
2.240-45	2.2404	49½	40½				2.2411	42¼	47¾			
2.245-50	2.2488	49¾	40¼	2.2489	45¾	44¼				2.2473	39¼	50¾

LEADS FROM 2.250 TO 2.400

Approximate Lead	Gear on Worm 28, 1st on Stud 64, 2nd on Stud 56, Gear on Screw 100	Angle to Set Spiral Head (Degrees)	Angle to Set Vertical Att. (Degrees)	Gear on Worm 40, 1st on Stud 86, 2nd on Stud 56, Gear on Screw 100	Angle to Set Spiral Head (Degrees)	Angle to Set Vertical Att. (Degrees)	Gear on Worm 40, 1st on Stud 56, 2nd on Stud 28, Gear on Screw 72	Angle to Set Spiral Head (Degrees)	Angle to Set Vertical Att. (Degrees)	Gear on Worm 24, 1st on Stud 56, 2nd on Stud 44, Gear on Screw 64	Angle to Set Spiral Head (Degrees)	Angle to Set Vertical Att. (Degrees)	Gear on Worm 24, 1st on Stud 86, 2nd on Stud 72, Gear on Screw 64	Angle to Set Spiral Head (Degrees)	Angle to Set Vertical Att. (Degrees)	Gear on Worm 28, 1st on Stud 56, 2nd on Stud 48, Gear on Screw 72	Angle to Set Spiral Head (Degrees)	Angle to Set Vertical Att. (Degrees)	Gear on Worm 48, 1st on Stud 44, 2nd on Stud 28, Gear on Screw 96	Angle to Set Spiral Head (Degrees)	Angle to Set Vertical Att. (Degrees)	Gear on Worm 44, 1st on Stud 56, 2nd on Stud 48, Gear on Screw 100	Angle to Set Spiral Head (Degrees)	Angle to Set Vertical Att. (Degrees)
2.250-55	2.2510	66¾	23¼	2.2500	59¾	30¼	2.2545	54¼	35¾							2.2520	42½	47½						
2.255-60	2.2551	67	23	2.2559	60	30				2.2571	50	40	2.2584	46	44				2.2594	39½	50½	2.2565	36¾	53¼
2.260-65	2.2634	67½	22½	2.2616	60¼	29¾	2.2615	54½	35½	2.2653	50¼	39¾				2.2625	42¾	47¼						
2.265-70	2.2679	67¾	22¼	2.2671	60½	29½	2.2686	54¾	35¼				2.2679	46¼	43¾							2.2697	37	53
2.270-75	2.2716	68	22	2.2729	61¾	28¼				2.2735	50½	39½				2.2732	43	47	2.2713	39¾	50¼			
2.275-80	2.2793	68½	21½	2.2781	61	29	2.2755	55	35				2.2773	46½	43½									
2.280-85	2.2834	68¾	21¼	2.2836	61¼	28¾	2.2825	55¼	34¾	2.2816	50¾	39¼				2.2839	43¼	46¾	2.2862	40	50	2.2827	37¼	52¾
2.285-90	2.2873	69	21	2.2890	61½	28½	2.2894	55½	34½	2.2896	51	39	2.2867	46¾	43¼									
2.290-95	2.2910	69¼	20¾	2.2948	61¾	28¼										2.2945	43½	46½						
2.295-00	2.2986	69¾	20¼				2.2961	55¾	34¼	2.2978	51¼	38¾	2.2961	47	43				2.2950	40¼	49¾	2.2959	37½	52½
2.300-05	2.3022	70	20	2.3001	62	28	2.3030	56	34							2.3049	43¾	46¼				2.3089	37¾	52¼
2.305-10	2.3095	70½	19½	2.3054	62¼	27¾	2.3097	56¼	33¾	2.3059	51½	38½	2.3055	47¼	42¾				2.3068	40½	49½			
2.310-15	2.3130	70¾	19¼	2.3104	62½	27½				2.3137	51¾	38¼												
2.315-20	2.3164	71	19	2.3158	62¾	27¼	2.3163	56½	33½				2.3148	47½	42½	2.3153	44	46	2.3185	40¾	49¼			
2.320-25	2.3200	71¼	18¾	2.3209	63	27	2.3239	56¾	33¼	2.3217	52	38	2.3239	47¾	42¼							2.3220	38	52
2.325-30	2.3267	71¾	18¼	2.3261	63¼	26¾				2.3299	52¼	37¾				2.3258	44¼	45¾						
2.330-35	2.3332	72¼	17¾	2.3311	63½	26½	2.3300	57	33				2.3331	48	42				2.3302	41	49	2.3348	38¼	51¾
2.335-40	2.3398	72¾	17¼	2.3364	63¾	26¼	2.3361	57¼	32¾	2.3376	52½	37½				2.3362	44½	45½						
2.340-45	2.3428	73	17	2.3411	64	26	2.3430	57½	32½				2.3423	48¼	41¾				2.3419	41¼	48¾			
2.345-50	2.3459	73¼	16¾	2.3461	64¼	25¾	2.3491	57¾	32¼	2.3454	52¾	37¼				2.3465	44¾	45¼				2.3476	38½	51½
2.350-55	2.3520	73¾	16¼	2.3511	64½	25½				2.3531	53	37	2.3514	48½	41½				2.3535	41½	48½			
2.355-60	2.3580	74½	15½	2.3560	64¾	25¼	2.3555	58	32							2.3570	45	45						
2.360-65	2.3638	74¾	15¼	2.3608	65	25	2.3620	58¼	31¾	2.3608	53¼	36¾	2.3604	48¾	41¼							2.3605	38¾	51¼
2.365-70	2.3665	75	15	2.3655	65¼	24¾	2.3685	58½	31½	2.3685	53½	36½	2.3694	49	41	2.3672	45¼	44¾	2.3651	41¾	48¼			
2.370-75	2.3720	75½	14½	2.3702	65½	24½	2.3748	58¾	31¼													2.3732	39	51
2.375-80	2.3798	76¼	13¾	2.3750	65¾	24¼				2.3761	53¾	36¼	2.3784	49¼	40¾	2.3773	45½	44½	2.3767	42	48			
2.380-85	2.3823	76½	13½	2.3842	66¼	23¾	2.3811	59	31	2.3837	54	36												
2.385-90	2.3871	77	13	2.3890	66½	23½	2.3872	59¼	30¾				2.3873	49½	40½	2.3875	45¾	44¼	2.3883	42¼	47¾	2.3861	39¼	50¾
2.390-95	2.3920	77½	12½	2.3929	66¾	23¼	2.3935	59½	30½	2.3912	54¼	35¾												
2.395-00	2.3965	78	12	2.3978	67	23	2.3996	59¾	30¼	2.3985	54½	35½	2.3962	49¾	40¼	2.3977	46	44	2.3997	42½	47½	2.3989	39½	50½

LEADS FROM 2.400 TO 2.550

Approximate Lead	Gear on Worm 40, 1st on Stud 86, 2nd on Stud 56, Gear on Screw 100	Angle to Set Spiral Head (Degrees)	Angle to Set Vertical Att. (Degrees)	Gear on Worm 40, 1st on Stud 56, 2nd on Stud 28, Gear on Screw 72	Angle to Set Spiral Head (Degrees)	Angle to Set Vertical Att. (Degrees)	Gear on Worm 24, 1st on Stud 56, 2nd on Stud 44, Gear on Screw 64	Angle to Set Spiral Head (Degrees)	Angle to Set Vertical Att. (Degrees)	Gear on Worm 24, 1st on Stud 86, 2nd on Stud 72, Gear on Screw 64	Angle to Set Spiral Head (Degrees)	Angle to Set Vertical Att. (Degrees)	Gear on Worm 28, 1st on Stud 56, 2nd on Stud 48, Gear on Screw 72	Angle to Set Spiral Head (Degrees)	Angle to Set Vertical Att. (Degrees)	Gear on Worm 48, 1st on Stud 44, 2nd on Stud 28, Gear on Screw 86	Angle to Set Spiral Head (Degrees)	Angle to Set Vertical Att. (Degrees)	Gear on Worm 44, 1st on Stud 56, 2nd on Stud 48, Gear on Screw 100	Angle to Set Spiral Head (Degrees)	Angle to Set Vertical Att. (Degrees)	Gear on Worm 72, 1st on Stud 100, 2nd on Stud 48, Gear on Screw 86	Angle to Set Spiral Head (Degrees)	Angle to Set Vertical Att. (Degrees)
2.400-05	2.4021	67¼	22¾																			2.4044	36¾	53¼
2.405-10	2.4065	67½	22½	2.4058	60	30	2.4060	54¾	35¼	2.4049	50	40	2.4079	46¼	43¾									
2.410-15	2.4111	67¾	22¼	2.4118	60¼	29¾	2.4135	55	35	2.4137	50¼	39¾				2.4113	42¾	47¼	2.4116	39¾	50¼			
2.415-20	2.4150	68	22	2.4179	60½	29½							2.4179	46½	43½							2.4184	37	53
2.420-25	2.4235	68½	21½	2.4238	60¾	29¼	2.4209	55¼	34¾	2.4224	50½	39½				2.4224	43	47	2.4242	40	50			
2.425-30	2.4278	68¾	21¼	2.4297	61	29	2.4282	55½	34½				2.4279	46¾	43¼									
2.430-35	2.4318	69	21							2.4311	50¾	39¼										2.4324	37¼	52¾
2.435-40	2.4359	69¼	20¾	2.4354	61¼	28¾	2.4354	55¾	34¼	2.4398	51	39	2.4378	47	43	2.4361	43¼	46¾	2.4368	40¼	49¾			
2.440-45	2.4438	69¾	20¼	2.4410	61½	28½	2.4426	56	34							2.4446	43½	46½						
2.445-50	2.4479	70	20	2.4471	61¾	28¼	2.4498	56¼	33¾	2.4484	51¼	38¾	2.4477	47¼	42¾				2.4494	40½	49½	2.4464	37½	52½
2.450-55	2.4515	70¼	19¾	2.4529	62	28																		
2.455-60	2.4554	70½	19½	2.4585	62¼	27¾	2.4569	56½	33½	2.4569	51½	38½	2.4576	47½	42½	2.4561	43¾	46¼						
2.460-65	2.4630	71	19	2.4640	62½	27½	2.4640	56¾	33¼										2.4619	40¾	49¼	2.4603	37¾	52¼
2.465-70	2.4666	71¼	18¾	2.4695	62¾	27¼				2.4654	51¾	38¼	2.4674	47¾	42¼	2.4672	44	46						
2.470-75	2.4702	71½	18½				2.4710	57	33	2.4739	52	38							2.4743	41	49	2.4741	38	52
2.475-80	2.4774	72	18	2.4750	63	27	2.4780	57¼	32¾				2.4771	48	42	2.4783	44¼	45¾						
2.480-85	2.4841	72½	17½	2.4808	63¼	26¾	2.4849	57½	32½	2.4823	52¼	37¾												
2.485-90	2.4878	72¾	17¼	2.4860	63½	26½							2.4868	48¼	41¾	2.4895	44½	45½	2.4867	41¼	48¾	2.4879	38¼	51¾
2.490-95	2.4942	73¼	16¾	2.4915	63¾	26¼	2.4918	57¾	32¼	2.4907	52½	37½												
2.495-00	2.4976	73½	16½	2.4969	64	26	2.4987	58	32	2.4990	52¾	37¼	2.4965	48½	41½				2.4990	41½	48½			
2.500-05	2.5040	74	16	2.5020	64¼	25¾										2.5005	44¾	45¼				2.5016	38½	51½
2.505-10	2.5071	74¼	15¾	2.5074	64½	25½	2.5055	58¼	31¾	2.5073	53	37	2.5061	48¾	41¼									
2.510-15	2.5102	74½	15½	2.5126	64¾	25¼	2.5122	58½	31½							2.5116	45	45	2.5113	41¾	48¼			
2.515-20	2.5191	75¼	14¾	2.5176	65	25	2.5189	58¾	31¼	2.5155	53¼	36¾	2.5157	49	41							2.5153	38¾	51¼
2.520-25	2.5220	75½	14½	2.5226	65¼	24¾				2.5237	53½	36½				2.5226	45¼	44¾	2.5236	42	48			
2.525-30	2.5276	76	14	2.5279	65½	24½	2.5256	59	31				2.5252	49¼	40¾							2.5290	39	51
2.530-35	2.5326	76½	13½	2.5329	65¾	24¼	2.5321	59¼	30¾	2.5318	53¾	36¼	2.5347	49½	40½	2.5335	45½	44½						
2.535-40	2.5379	77	13	2.5380	66	24	2.5387	59½	30½	2.5399	54	36							2.5358	42¼	47¾			
2.540-45	2.5430	77½	12½	2.5426	66¼	23¾							2.5441	49¾	40¼	2.5442	45¾	44¼				2.5426	39¼	50¼
2.545-50	2.5478	78	12	2.5478	66½	23½	2.5452	59¾	30¼	2.5484	54¼	35¾							2.5480	42½	37½			

LEADS FROM 2.550 TO 2.700

Approximate Lead	Gear on Worm 40, 1st on Stud 56, 2nd on Stud 28, Gear on Screw 72	Angle to Set Spiral Head (Degrees)	Angle to Set Vertical Att. (Degrees)	Gear on Worm 24, 1st on Stud 56, 2nd on Stud 44, Gear on Screw 64	Angle to Set Spiral Head (Degrees)	Angle to Set Vertical Att. (Degrees)	Gear on Worm 24, 1st on Stud 86, 2nd on Stud 72, Gear on Screw 64	Angle to Set Spiral Head (Degrees)	Angle to Set Vertical Att. (Degrees)	Gear on Worm 28, 1st on Stud 56, 2nd on Stud 48, Gear on Screw 72	Angle to Set Spiral Head (Degrees)	Angle to Set Vertical Att. (Degrees)	Gear on Worm 48, 1st on Stud 44, 2nd on Stud 28, Gear on Screw 86	Angle to Set Spiral Head (Degrees)	Angle to Set Vertical Att. (Degrees)	Gear on Worm 44, 1st on Stud 56, 2nd on Stud 48, Gear on Screw 100	Angle to Set Spiral Head (Degrees)	Angle to Set Vertical Att. (Degrees)	Gear on Worm 72, 1st on Stud 100, 2nd on Stud 48, Gear on Screw 86	Angle to Set Spiral Head (Degrees)	Angle to Set Vertical Att. (Degrees)	Gear on Worm 64, 1st on Stud 48, 2nd on Stud 32, Gear on Screw 100	Angle to Set Spiral Head (Degrees)	Angle to Set Vertical Att. (Degrees)
2.550-55	2.5522	66¾	23¼	2.5517	60	30				2.5534	50	40										2.5528	36¾	53¼
2.555-60	2.5569	67	23	2.5581	60¼	29¾	2.5560	54½	35½	2.5628	50¼	39¾	2.5557	46	44				2.5561	39½	50½			
2.560-65	2.5617	67¼	22¾	2.5645	60½	29½	2.5639	54¾	35¼							2.5600	42¾	47¼						
2.565-70	2.5663	67½	22½										2.5658	46¼	43¾				2.5696	39¾	50¼	2.5678	37	53
2.570-75	2.5709	67¾	22¼	2.5708	60¾	29¼	2.5717	55	35	2.5720	50½	39½				2.5721	43	47						
2.575-80	2.5755	68	22	2.5770	61	29	2.5796	55¼	34¾				2.5762	46½	43½									
2.580-85	2.5845	68½	21½	2.5832	61¼	28¾				2.5812	50¾	39¼							2.5831	40	50	2.5826	37¼	52¾
2.585-90	2.5890	68¾	21¼	2.5894	61½	28½	2.5874	55½	34½				2.5871	46¾	43¼	2.5841	43¼	46¾						
2.590-95	2.5932	69	21							2.5904	51	39												
2.595-00	2.5976	69¼	20¾	2.5955	61¾	28¼	2.5951	55¾	34¼	2.5995	51¼	38¾	2.5976	47	43	2.5961	43½	46½	2.5965	40¼	49¾	2.5974	37½	52½
2.600-05	2.6019	69½	20½	2.6016	62	28	2.6029	56	34															
2.605-10	2.6060	69¾	20¼	2.6076	62¼	27¾				2.6086	51½	38½	2.6082	47¼	42¾	2.6080	43¾	46¼	2.5099	40½	49½	2.6122	37¾	52¼
2.610-15	2.6143	70¼	19¾	2.6135	62½	27½	2.6104	56¼	33¾															
2.615-20	2.6185	70½	19½	2.6194	62¾	27¼	2.6180	56½	33½	2.6177	51¾	38¼	2.6188	47½	42½	2.6199	44	46						
2.620-25	2.6227	70¾	19¼	2.6253	63	27													2.6232	40¾	49¼			
2.625-30	2.6265	71	19				2.6256	56¾	33¼	2.6267	52	38	2.6290	47¾	42¼							2.6268	38	52
2.630-35	2.6342	71½	18½	2.6311	63¼	26¾	2.6330	57	33							2.6317	44¼	45¾						
2.635-40	2.6381	71¾	18¼	2.6369	63½	26½				2.6356	52¼	37¾							2.6364	41	49			
2.640-45	2.6420	72	18	2.6426	63¾	26¼	2.6405	57¼	32¾	2.6445	52½	37½	2.6460	48	42	2.6434	44½	45½				2.6415	38¼	51¾
2.645-50	2.6457	72¼	17¾	2.6482	64	26	2.6479	57½	32½										2.6496	41¼	48¾			
2.650-55	2.6530	72¾	17¼	2.6538	64¼	25¾				2.6533	52¾	37¼	2.6500	48¼	41¾	2.6551	44¾	45¼						
2.655-60	2.6565	73	17	2.6594	64½	25½	2.6552	57¾	32¼													2.6560	38½	51½
2.660-65	2.6636	73½	16½	2.6649	64¾	25¼	2.6625	58	32	2.6620	53	37	2.6601	48½	41½				2.6628	41½	48½			
2.665-70	2.6670	73¾	16¼				2.6697	58¼	31¾							2.6668	45	45						
2.670-75	2.6737	74¼	15¾	2.6704	65	25				2.6708	53¼	36¾	2.6706	48¾	41¼							2.6705	38¾	51¼
2.675-80	2.6770	74½	15½	2.6758	65¼	24¾	2.6769	58½	31½	2.6795	53½	36½				2.6784	45¼	44¾	2.6759	41¾	48¼			
2.680-85	2.6831	75	15	2.6811	65½	24½	2.6840	58¾	31¼				2.6801	49	41									
2.685-90	2.6892	75½	14½	2.6864	65¾	24¼				2.6880	53¾	36¼							2.6890	42	48	2.6851	39	51
2.690-95	2.6922	75¾	14¼	2.6917	66	24	2.6911	59	31							2.6900	45½	44½						
2.695-00	2.6952	76	14	2.6969	66¼	23¾	2.6981	59¼	30¾	2.6968	54	36	2.6989	49¼	40¾							2.6996	39¼	50¾

Approximate Lead	Gear on Worm 24, 1st on Stud 56, 2nd on Stud 44, Gear on Screw 64	Angle to Set Spiral Head (Degrees)	Angle to Set Vertical Att. (Degrees)	Gear on Worm 24, 1st on Stud 86, 2nd on Stud 72, Gear on Screw 64	Angle to Set Spiral Head (Degrees)	Angle to Set Vertical Att. (Degrees)	Gear on Worm 28, 1st on Stud 56, 2nd on Stud 48, Gear on Screw 72	Angle to Set Spiral Head (Degrees)	Angle to Set Vertical Att. (Degrees)	Gear on Worm 48, 1st on Stud 44, 2nd on Stud 28, Gear on Screw 86	Angle to Set Spiral Head (Degrees)	Angle to Set Vertical Att. (Degrees)
2.700-05	2.7020	66½	23½							2.7009	49½	40½
2.705-10	2.7071	66¾	23¼	2.7051	59½	30½	2.7051	54¼	35¾			
2.710-15	2.7122	67	23	2.7120	59¾	30¼	2.7135	54½	35½	2.7110	49¾	40¼
2.715-20	2.7172	67¼	22¾	2.7189	60	30						
2.720-25	2.7222	67½	22½				2.7220	54¾	35¼	2.7210	50	40
2.725-30	2.7271	67¾	22¼	2.7257	60¼	29¾						
2.730-35	2.7319	68	22	2.7325	60½	29½	2.7304	55	35	2.7310	50¼	39¾
2.735-40	2.7367	68¼	21¾	2.7392	60¾	29¼	2.7387	55¼	34¾			
2.740-45	2.7414	68½	21½							2.7407	50½	39½
2.745-50	2.7461	68¾	21¼	2.7459	61	29	2.7470	55½	34½			
2.750-55	2.7507	69	21	2.7525	61¼	28¾				2.7505	50¾	39¼
2.755-60	2.7598	69½	20½	2.7591	61½	28½	2.7552	55¾	34¼			
2.760-65	2.7643	69¾	20¼				2.7634	56	34	2.7601	51	39
2.765-70	2.7687	70	20	2.7656	61¾	28¼						
2.770-75	2.7731	70¼	19¾	2.7721	62	28	2.7715	56¼	33¾	2.7700	51¼	38¾
2.775-80	2.7774	70½	19½	2.7785	62¼	27¾	2.7796	56½	33½	2.7797	51½	38½
2.780-85	2.7817	70¾	19¼									
2.785-90	2.7859	71	19	2.7848	62½	27½	2.7876	56¾	33¼	2.7891	51¾	38¼
2.790-95	2.7942	71½	18½	2.7911	62¾	27¼						
2.795-00	2.7982	71¾	18¼	2.7974	63	27	2.7955	57	33	2.7989	52	38
2.800-05	2.8022	72	18	2.8036	63¼	26¾	2.8034	57¼	32¾			
2.805-10	2.8062	72¼	17¾	2.8097	63½	26½				2.8087	52¼	37¾
2.810-15	2.8139	72¾	17¼				2.8113	57½	32½			
2.815-20	2.8177	73	17	2.8158	63¾	26¼	2.8191	57¾	32¼	2.8180	52½	37½
2.820-25	2.8214	73¼	16¾	2.8218	64	26						
2.825-30	2.8251	73½	16¼	2.8278	64¼	25¾	2.8268	58	32	2.8275	52¾	37¼
2.830-35	2.8323	74	16	2.8337	64½	25½	2.8345	58¼	31¾			
2.835-40	2.8393	74½	15½	2.8396						2.8369	53	37
2.840-45	2.8427	74¾	15¼				2.8421	58½	31½			
2.845-50	2.8461	75	15	2.8454	65	25	2.8497	58¾	31¼	2.8461	53¼	36¾

Approximate Lead	Gear on Worm 44, 1st on Stud 56, 2nd on Stud 48, Gear on Screw 100	Angle to Set Spiral Head (Degrees)	Angle to Set Vertical Att. (Degrees)	Gear on Worm 72, 1st on Stud 100, 2nd on Stud 48, Gear on Screw 86	Angle to Set Spiral Head (Degrees)	Angle to Set Vertical Att. (Degrees)	Gear on Worm 64, 1st on Stud 48, 2nd on Stud 32, Gear on Screw 100	Angle to Set Spiral Head (Degrees)	Angle to Set Vertical Att. (Degrees)	Gear on Worm 56, 1st on Stud 48, 2nd on Stud 28, Gear on Screw 72	Angle to Set Spiral Head (Degrees)	Angle to Set Vertical Att. (Degrees)
2.700-05	2.7015	45¾	44¼	2.7020	42¼	47¾						
2.705-10												
2.710-15	2.7129	46	44	2.7149	42½	47½	2.7141	39½	50½	2.7144	36¾	53¼
2.715-20												
2.720-25	2.7243	46¼	43¾									
2.725-30				2.7278	42¾	47¼	2.7282	39¾	50¼	2.7303	37	53
2.730-35												
2.735-40	2.7357	46½	43½									
2.740-45				2.7407	43	47	2.7427	40	50			
2.745-50	2.7470	46¾	43¼							2.7460	37¼	52¾
2.750-55												
2.755-60	2.7582	47	43	2.7535	43¼	46¾	2.7567	40¼	49¾			
2.760-65										2.7618	37½	52½
2.765-70	2.7694	47¼	42¾	2.7662	43½	46½						
2.770-75							2.7710	40½	49½			
2.775-80				2.7789	43¾	46¼				2.7775	37¾	52¼
2.780-85	2.7806	47½	42½									
2.785-90							2.7851	40¾	49¼			
2.790-95	2.7917	47¾	42¼	2.7915	44	46				2.7931	38	52
2.795-00							2.7991	41	49			
2.800-05	2.8027	48	42	2.8041	44¼	45¾				2.8086	38¼	51¾
2.805-10												
2.810-15	2.8137	48¼	41¾				2.8132	41¼	48¾			
2.815-20				2.8166	44½	45½						
2.820-25	2.8246	48½	41½							2.8240	38½	51½
2.825-30				2.8291	44¾	45¼	2.8272	41½	48½			
2.830-35												
2.835-40	2.8355	48¾	41¼							2.8396	38¾	51¼
2.840-45				2.8416	45	45	2.8411	41¾	48¼			
2.845-50	2.8463	49	41									

LEADS FROM 2.850 TO 3.000

Approximate Lead	Gear on Worm 24, 1st on Stud 86, 2nd on Stud 72, Gear on Screw 64	Angle to Set Spiral Head (Degrees)	Angle to Set Vertical Att. (Degrees)	Gear on Worm 28, 1st on Stud 56, 2nd on Stud 48, Gear on Screw 72	Angle to Set Spiral Head (Degrees)	Angle to Set Vertical Att. (Degrees)	Gear on Worm 48, 1st on Stud 44, 2nd on Stud 28, Gear on Screw 86	Angle to Set Spiral Head (Degrees)	Angle to Set Vertical Att. (Degrees)	Gear on Worm 44, 1st on Stud 56, 2nd on Stud 48, Gear on Screw 100	Angle to Set Spiral Head (Degrees)	Angle to Set Vertical Att. (Degrees)
2.850-55	2.8512	65¼	24¾									
2.855-60	2.8569	65½	24½	2.8572	59	31	2.8553	53½	36½	2.8571	49¼	40¾
2.860-65	2.8625	65¾	24¼	2.8647	59¼	30¾	2.8645	53¾	36¼			
2.865-70	2.8682	66	24							2.8678	49½	40½
2.870-75	2.8736	66¼	23¾	2.8721	59½	30½	2.8737	54	36			
2.875-80	2.8792	66½	23½	2.8795	59¾	30¼				2.8785	49¾	40¼
2.880-85							2.8825	54¼	35¾			
2.885-90	2.8846	66¾	23¼	2.8868	60	30				2.8891	50	40
2.890-95	2.8900	67	23	2.8940	60¼	29¾	2.8916	54½	35½			
2.895-00	2.8953	67¼	22¾							2.8996	50¼	39¾
2.900-05	2.9006	67½	22½	2.9012	60½	29½	2.9005	54¾	35¼			
2.905-10	2.9058	67¾	22¼	2.9083	60¾	29¼	2.9095	55	35			
2.910-15	2.9109	68	22							2.9101	50½	39½
2.915-20	2.9160	68¼	21¾	2.9154	61	29	2.9185	55¼	34¾			
2.920-25	2.9211	68½	21½	2.9224	61¼	28¾				2.9206	50¾	39¼
2.925-30	2.9261	68¾	21¼	2.9294	61½	28½	2.9273	55½	34½			
2.930-35	2.9310	69	21							2.9310	51	39
2.935-40	2.9359	69¼	20¾	2.9363	61¾	28¼	2.9360	55¾	34¼			
2.940-45	2.9407	69½	20½	2.9432	62	28	2.9446	56	34	2.9413	51¼	38¾
2.945-50	2.9455	69¾	20¼									
2.950-55	2.9549	70¼	19¾	2.9500	62¼	27¾	2.9532	56¼	33¾	2.9516	51½	38½
2.955-60	2.9595	70½	19½	2.9567	62½	27½						
2.960-65	2.9640	70¾	19¼	2.9634	62¾	27¼	2.9618	56½	33½	2.9618	51¾	38¼
2.965-70	2.9685	71	19									
2.970-75	2.9729	71¼	18¾	2.9700	63	27	2.9704	56¾	33¼	2.9720	52	38
2.975-80	2.9773	71½	18½	2.9766	63¼	26¾	2.9789	57	33			
2.980-85	2.9816	71¾	18¼	2.9831	63½	26½				2.9821	52¼	37¾
2.985-90	2.9859	72	18	2.9898	63¾	26¼	2.9874	57¼	32¾			
2.990-95	2.9942	72½	17½							2.9921	52½	37½
2.995-00	2.9983	72¾	17¼	2.9960	64	26	2.9958	57½	32½			

Approximate Lead	Gear on Worm 72, 1st on Stud 100, 2nd on Stud 48, Gear on Screw 86	Angle to Set Spiral Head (Degrees)	Angle to Set Vertical Att. (Degrees)	Gear on Worm 64, 1st on Stud 48, 2nd on Stud 32, Gear on Screw 100	Angle to Set Spiral Head (Degrees)	Angle to Set Vertical Att. (Degrees)	Gear on Worm 56, 1st on Stud 48, 2nd on Stud 28, Gear on Screw 72	Angle to Set Spiral Head (Degrees)	Angle to Set Vertical Att. (Degrees)	Gear on Worm 56, 1st on Stud 64, 2nd on Stud 40, Gear on Screw 72	Angle to Set Spiral Head (Degrees)	Angle to Set Vertical Att. (Degrees)
2.850-55	2.8540	45¼	44¾									
2.855-60				2.8541	42	48	2.8552	39	51	2.8573	36	54
2.860-65												
2.865-70	2.8663	45½	44½	2.8689	42¼	47¾						
2.870-75							2.8705	39¼	50¾	2.8744	36¼	53¾
2.875-80	2.8785	45¾	44¼									
2.880-85				2.8825	42½	47½						
2.885-90							2.8858	39½	50½			
2.890-95	2.8907	46	44							2.8915	36½	53½
2.895-00				2.8962	42¾	47¼						
2.900-05	2.9029	46¼	43¾				2.9011	39¾	50¼			
2.905-10				2.9099	43	47				2.9085	36¾	53¼
2.910-15												
2.915-20	2.9150	46½	43½				2.9163	40	50			
2.920-25												
2.925-30	2.9270	46¾	43¼	2.9234	43¼	46¾				2.9255	37	53
2.930-35							2.9314	40¼	49¾			
2.935-40	2.9390	47	43	2.9370	43½	46½						
2.940-45										2.9424	37¼	52¾
2.945-50							2.9465	40½	49½			
2.950-55	2.9509	47¼	42¾	2.9503	43¾	46¼						
2.955-60										2.9592	37½	52½
2.960-65	2.9628	47½	42½	2.9638	44	46	2.9615	40¾	49¼			
2.965-70												
2.970-75	2.9746	47¾	42¼									
2.975-80				2.9771	44¼	45¾	2.9765	41	49	2.9761	37¾	52¼
2.980-85												
2.985-90	2.9864	48	42									
2.990-95				2.9905	44½	45½	2.9915	41¼	48¾	2.9929	38	52
2.995-00	2.9982	48¼	41¾									

LEADS FROM 3.000 TO 3.150

Approximate Lead	Gear on Worm 28, 1st on Stud 56, 2nd on Stud 48, Gear on Screw 72	Angle to Set Spiral Head	Angle to Set Vertical Att.	Gear on Worm 48, 1st on Stud 44, 2nd on Stud 28, Gear on Screw 86	Angle to Set Spiral Head	Angle to Set Vertical Att.	Gear on Worm 44, 1st on Stud 56, 2nd on Stud 48, Gear on Screw 100	Angle to Set Spiral Head	Angle to Set Vertical Att.	Gear on Worm 72, 1st on Stud 100, 2nd on Stud 48, Gear on Screw 86	Angle to Set Spiral Head	Angle to Set Vertical Att.	Gear on Worm 64, 1st on Stud 48, 2nd on Stud 32, Gear on Screw 100	Angle to Set Spiral Head	Angle to Set Vertical Att.	Gear on Worm 56, 1st on Stud 48, 2nd on Stud 28, Gear on Screw 72	Angle to Set Spiral Head	Angle to Set Vertical Att.	Gear on Worm 56, 1st on Stud 64, 2nd on Stud 40, Gear on Screw 72	Angle to Set Spiral Head	Angle to Set Vertical Att.	Gear on Worm 56, 1st on Stud 48, 2nd on Stud 44, Gear on Screw 100	Angle to Set Spiral Head	Angle to Set Vertical Att.
		Degrees			Degrees			Degrees			Degrees			Degrees			Degrees			Degrees			Degrees	
3.000-05	3.002	64¼	25¾	3.004	57¾	32¼	3.002	52¾	37¼				3.004	44¾	45¼									
3.005-10	3.009	64½	25½													3.006	41½	48½	3.009	38¼	51¾			
3.010-15				3.012	58	32	3.012	53	37	3.010	48½	41½												
3.015-20	3.015	64¾	25¼										3.017	45	45							3.017	36	54
3.020-25	3.021	65	25	3.020	58¼	31¾	3.021	53¼	36¾	3.022	48¾	41¼				3.021	41¾	48¼						
3.025-30	3.027	65¼	24¾	3.028	58½	31½													3.026	38½	51½			
3.030-35	3.033	65½	24½				3.031	53½	36½	3.033	49	41	3.030	45¼	44¾									
3.035-40	3.039	65¾	24¼	3.036	58¾	31¼										3.036	42	48				3.035	36¼	53¾
3.040-45				3.044	59	31	3.041	53¾	36¼				3.043	45½	44½				3.042	38¾	51¼			
3.045-50	3.045	66	24							3.045	49¼	40¾												
3.050-55	3.051	66¼	23¾	3.052	59¼	30¾	3.051	54	36							3.050	42¼	47¾				3.053	36½	53½
3.055-60	3.057	66½	23½							3.056	49½	40½	3.056	45¾	44¼				3.059	39	51			
3.060-65	3.062	66¾	23¼	3.060	59½	30½	3.060	54¼	35¾															
3.065-70	3.068	67	23	3.068	59¾	30¼				3.067	49¾	40¼	3.069	46	44	3.065	42½	47½						
3.070-75	3.074	67¼	22¾				3.070	54½	35½													3.071	36¾	53¼
3.075-80				3.076	60	30	3.079	54¾	35¼	3.079	50	40							3.075	39¼	50¾			
3.080-85	3.080	67½	22½	3.084	60¼	29¾							3.082	46¼	43¾	3.080	42¾	47¼						
3.085-90	3.085	67¾	22¼				3.089	55	35													3.089	37	53
3.090-95	3.090	68	22	3.091	60½	29½				3.090	50¼	39¾				3.094	43	47	3.091	39½	50½			
3.095-00	3.096	68¼	21¾	3.099	60¾	29¼	3.098	55¼	34¾				3.095	46½	43½									
3.100-05	3.101	68½	21½							3.101	50½	39½												
3.105-10	3.107	68¾	21¼	3.107	61	29	3.108	55½	34½				3.108	46¾	43¼	3.109	43¼	46¾	3.108	39¾	50¼	3.107	37¼	52¾
3.110-15	3.112	69	21	3.114	61¼	28¾				3.112	50¾	39¼												
3.115-20	3.117	69¼	20¾				3.117	55¾	34¼															
3.120-25	3.122	69½	20½	3.121	61½	28½				3.123	51	39	3.120	47	43	3.123	43½	46½	3.124	40	50			
3.125-30	3.127	69¾	20¼	3.128	61¾	28¼	3.126	56	34													3.125	37½	52½
3.130-35	3.132	70	20							3.134	51¼	38¾	3.133	47¼	42¾									
3.135-40	3.137	70¼	19¾	3.136	62	28	3.135	56¼	33¾							3.137	43¾	46¼						
3.140-45	3.142	70½	19½	3.144	62¼	27¾	3.144	56½	33½										3.141	40¼	49¾	3.142	37¾	52¼
3.145-50	3.147	70¾	19¼							3.145	51½	38½	3.146	47½	42½									

LEADS FROM 3.150 TO 3.300

Approximate Lead	Gear on Worm 48, 1st on Stud 44, 2nd on Stud 28, Gear on Screw 86	Angle to Set Spiral Head (Degrees)	Angle to Set Vertical Att. (Degrees)	Gear on Worm 44, 1st on Stud 56, 2nd on Stud 48, Gear on Screw 100	Angle to Set Spiral Head (Degrees)	Angle to Set Vertical Att. (Degrees)	Gear on Worm 72, 1st on Stud 100, 2nd on Stud 48, Gear on Screw 86	Angle to Set Spiral Head (Degrees)	Angle to Set Vertical Att. (Degrees)	Gear on Worm 64, 1st on Stud 48, 2nd on Stud 32, Gear on Screw 100	Angle to Set Spiral Head (Degrees)	Angle to Set Vertical Att. (Degrees)	Gear on Worm 56, 1st on Stud 48, 2nd on Stud 28, Gear on Screw 72	Angle to Set Spiral Head (Degrees)	Angle to Set Vertical Att. (Degrees)	Gear on Worm 56, 1st on Stud 64, 2nd on Stud 40, Gear on Screw 72	Angle to Set Spiral Head (Degrees)	Angle to Set Vertical Att. (Degrees)	Gear on Worm 56, 1st on Stud 48, 2nd on Stud 44, Gear on Screw 100	Angle to Set Spiral Head (Degrees)	Angle to Set Vertical Att. (Degrees)	Gear on Worm 86, 1st on Stud 44, 2nd on Stud 28, Gear on Screw 100	Angle to Set Spiral Head (Degrees)	Angle to Set Vertical Att. (Degrees)
3.150-55	3.150	62½	27½										3.151	44	46									
3.155-60	3.158	62¾	27¼	3.155	56¾	33¼	3.156	51¾	38¼	3.158	47¾	42¼				3.157	40½	49½						
3.160-65				3.163	57	33													3.160	38	52	3.159	35¼	54¾
3.165-70	3.165	63	27				3.167	52	38				3.166	44¼	45¾									
3.170-75	3.172	63¼	26¾	3.171	57¼	32¾				3.171	48	42				3.173	40¾	49¼						
3.175-80	3.179	63½	26½				3.178	52¼	37¾													3.178	35½	54½
3.180-85				3.180	57½	32½				3.183	48¼	41¾	3.180	44½	45½				3.178	38¼	51¾			
3.185-90	3.186	63¾	26¼	3.189	57¾	32¼	3.189	52½	37½															
3.190-95	3.193	64	26										3.194	44¾	45¼	3.189	41	49						
3.195-00				3.198	58	32	3.199	52¾	37¼	3.195	48½	41½							3.195	38½	51½	3.197	35¾	54¼
3.200-05	3.200	64¼	25¾																					
3.205-10	3.206	64½	25½	3.206	58¼	31¾				3.208	48¾	41¼	3.208	45	45	3.205	41¼	48¾						
3.210-15	3.213	64¾	25¼				3.210	53	37										3.213	38¾	51¼	3.217	36	54
3.215-20	3.219	65	25	3.215	58½	31½																		
3.220-25				3.224	58¾	31¼	3.220	53¼	36¾	3.220	49	41	3.222	45¼	44¾	3.221	41½	48½						
3.225-30	3.226	65¼	24¾																					
3.230-35	3.232	65½	24½	3.232	59	31	3.231	53½	36½	3.232	49¼	40¾							3.230	39	51			
3.235-40	3.239	65¾	24¼										3.236	45½	44½	3.237	41¾	48¼				3.236	36¼	53¾
3.240-45				3.241	59¼	30¾	3.241	53¾	36¼	3.244	49½	40½												
3.245-50	3.245	66	24	3.249	59½	30½													3.248	39¼	50¾			
3.250-55	3.251	66¼	23¾				3.251	54	36				3.250	45¾	44¼	3.253	42	48						
3.255-60	3.257	66½	23½	3.257	59¾	30¼				3.257	49¾	40¼										3.255	36½	53½
3.260-65	3.264	66¾	23¼				3.262	54¼	35¾				3.264	46	44									
3.265-70				3.266	60	30				3.269	50	40				3.269	42¼	47¾	3.265	39½	50½			
3.270-75	3.270	67	23	3.274	60¼	29¾	3.272	54½	35½													3.274	36¾	53¼
3.275-80	3.276	67¼	22¾										3.277	46¼	43¾									
3.280-85	3.282	67½	22½	3.282	60½	29½	3.282	54¾	35¼	3.280	50¼	39¾				3.284	42½	47½	3.282	39¾	50¼			
3.285-90	3.288	67¾	22¼																					
3.290-95	3.293	68	22	3.290	60¾	29¼	3.292	55	35	3.292	50½	39½	3.291	46½	43½							3.294	37	53
3.295-00	3.299	68¼	21¾	3.298	61	29										3.299	42¾	47¼	3.299	40	50			

Approximate Lead	Gear on Worm 44, 1st on Stud 56, 2nd on Stud 48, Gear on Screw 100	Angle to Set Spiral Head (Degrees)	Angle to Set Vertical Att. (Degrees)	Gear on Worm 72, 1st on Stud 100, 2nd on Stud 48, Gear on Screw 86	Angle to Set Spiral Head (Degrees)	Angle to Set Vertical Att. (Degrees)	Gear on Worm 64, 1st on Stud 48, 2nd on Stud 32, Gear on Screw 100	Angle to Set Spiral Head (Degrees)	Angle to Set Vertical Att. (Degrees)	Gear on Worm 56, 1st on Stud 48, 2nd on Stud 28, Gear on Screw 72	Angle to Set Spiral Head (Degrees)	Angle to Set Vertical Att. (Degrees)	Gear on Worm 56, 1st on Stud 64, 2nd on Stud 40, Gear on Screw 72	Angle to Set Spiral Head (Degrees)	Angle to Set Vertical Att. (Degrees)	Gear on Worm 56, 1st on Stud 48, 2nd on Stud 44, Gear on Screw 100	Angle to Set Spiral Head (Degrees)	Angle to Set Vertical Att. (Degrees)	Gear on Worm 86, 1st on Stud 44, 2nd on Stud 28, Gear on Screw 100	Angle to Set Spiral Head (Degrees)	Angle to Set Vertical Att. (Degrees)	Gear on Worm 100, 1st on Stud 48, 2nd on Stud 24, Gear on Screw 86	Angle to Set Spiral Head (Degrees)	Angle to Set Vertical Att. (Degrees)
3.300-10	3.306	61¼	28¾	3.302	55¼	34¾	3.304	50¾	39¼	3.304	46¾	43¼												
3.310-20	3.314	61½	28½	3.312	55½	34½	3.316	51	39	3.318	47	43	3.315	43	47	3.316	40¼	49¾	3.313	37¼	52¾	3.314	34¾	55¼
3.320-30	3.322	61¾	28¼	3.322	55¾	34¼	3.327	51¼	38¾															
3.330-40	3.330	62	28	3.332	56	34	3.339	51½	38½	3.331	47¼	42¾	3.331	43¼	46¾	3.333	40½	49½	3.332	37½	52½	3.335	35	55
3.340-50	3.345	62½	27½	3.341	56¼	33¾				3.345	47½	42½	3.346	43½	46½									
3.350-60	3.352	62¾	27¼	3.351	56½	33½	3.351	51¾	38¼	3.358	47¾	42¼				3.350	40¾	49¼	3.351	37¾	52¼	3.356	35¼	54¾
3.360-70	3.360	63	27	3.361	56¾	33¼	3.362	52	38				3.361	43¾	46¼	3.368	41	49	3.369	38	52			
3.370-80	3.375	63½	26½	3.371	57	33	3.374	52¼	37¾	3.372	48	42	3.376	44	46							3.376	35½	54½
3.380-90	3.382	63¾	26¼	3.380	57¼	32¾	3.385	52½	37½	3.385	48¼	41¾				3.384	41¼	48¾	3.388	38¼	51¾			
3.390-00	3.390	64	26	3.390	57½	32½	3.397	52¾	37¼	3.398	48½	41½	3.392	44¼	45¾							3.397	35¾	54¼
3.400-10	3.404	64½	25½	3.408	58	32	3.408	53	37				3.407	44½	45½	3.401	41½	48½	3.407	38½	51½			
3.410-20	3.418	65	25	3.417	58¼	31¾	3.419	53¼	36¾	3.411	48¾	41¼				3.418	41¾	48¼				3.418	36	54
3.420-30	3.424	65¼	24¾	3.427	58½	31½				3.424	49	41	3.422	44¾	45¼				3.425	38¾	51¼			
3.430-40	3.431	65½	24½	3.436	58¾	31¼	3.430	53½	36½	3.437	49¼	41¾	3.437	45	45	3.435	42	48				3.438	36¼	53¾
3.440-50	3.445	66	24	3.445	59	31	3.441	53¾	36¼										3.444	39	51			
3.450-60	3.458	66½	23½	3.454	59¼	30¾	3.452	54	36	3.450	49½	41½	3.452	45¼	44¾	3.451	42¼	47¾				3.458	36½	53½
3.460-70	3.469	66¾	23¼	3.462	59½	30½	3.463	54¼	35¾	3.463	49¾	41¼	3.467	45½	44½	3.468	42½	47½	3.463	39¼	50¾			
3.470-80	3.470	67	23	3.471	59¾	30¼	3.473	54½	35½	3.475	50	40										3.478	36¾	53¼
3.480-90	3.484	67½	22½	3.480	60	30	3.484	54¾	35¼	3.488	50¼	39¾	3.482	45¾	44¼	3.485	42¾	47¼	3.481	39½	50½			
3.490-00	3.496	68	22	3.498	60½	29½	3.495	55	35				3.496	46	44				3.499	39¾	50¼	3.499	37	53
3.500-10	3.508	68½	21½	3.507	60¾	29¼	3.506	55¼	34¾	3.500	50½	39½				3.500	43	47						
3.510-20	3.514	68¾	21¼	3.515	61	29	3.516	55½	34½	3.513	50¾	39¼	3.511	46¼	43¾	3.517	43¼	46¾	3.518	40	50	3.519	37¼	52¾
3.520-30	3.520	69	21	3.523	61¼	28¾	3.527	55¾	34¼	3.525	51	39	3.526	46½	43½									
3.530-40	3.532	69½	20½	3.531	61½	28½	3.537	56	34	3.538	51¼	38¾				3.533	43½	46½	3.536	40¼	49¾	3.539	37½	52½
3.540-50	3.543	70	20	3.549	62	28	3.547	56¼	33¾				3.540	46¾	43¼	3.549	43¾	46¼						
3.550-60	3.544	70½	19½	3.557	62¼	27¾	3.558	56½	33½	3.550	51½	38½	3.555	47	43				3.554	40½	49½	3.559	37¾	52¼
3.560-70	3.565	71	19	3.565	62½	27½	3.568	56¾	33¼	3.562	51¾	38¼	3.569	47¼	42¾	3.565	44	46						
3.570-80	3.576	71½	18½	3.573	62¾	27¼	3.579	57	33	3.575	52	38							3.572	40¾	49¼	3.574	38	52
3.580-90	3.586	72	18	3.581	63	27	3.588	57¼	32¾	3.588	52¼	37¾	3.584	47½	42½	3.582	44¼	45¾						
3.590-00	3.591	72¼	17¾	3.596	63½	26½	3.599	57½	32½				3.597	47¾	42¼	3.597	44½	45½	3.590	41	49	3.599	38¼	51¾

LEADS FROM 3.600 TO 3.900

Approximate Lead	Gear on Worm 72 1st on Stud 100 2nd on Stud 48 Gear on Screw 86	Angle to Set Spiral Head (Degrees)	Angle to Set Vertical Att. (Degrees)	Gear on Worm 64 1st on Stud 48 2nd on Stud 32 Gear on Screw 100	Angle to Set Spiral Head (Degrees)	Angle to Set Vertical Att. (Degrees)	Gear on Worm 56 1st on Stud 48 2nd on Stud 28 Gear on Screw 72	Angle to Set Spiral Head (Degrees)	Angle to Set Vertical Att. (Degrees)	Gear on Worm 56 1st on Stud 64 2nd on Stud 40 Gear on Screw 72	Angle to Set Spiral Head (Degrees)	Angle to Set Vertical Att. (Degrees)
3.600-10	3.605	63¾	26¼	3.608	57¾	32¼	3.600	52½	37½			
3.610-20	3.612	64	26	3.618	58	32	3.612	52¾	37¼	3.613	48	42
3.620-30	3.620	64¼	25¾	3.628	58¼	31¾	3.623	53	37	3.627	48¼	41¾
3.630-40	3.635	64¾	25¼	3.638	58½	31½	3.635	53¼	36¾			
3.640-50	3.642	65	25	3.647	58¾	31¼	3.647	53½	36½	3.641	48½	41½
3.650-60	3.657	65½	24½	3.657	59	31	3.659	53¾	36¼	3.655	48¾	41¼
3.660-70	3.664	65¾	24¼	3.667	59¼	30¾				3.669	49	41
3.670-80	3.672	66	24	3.676	59½	30½	3.670	54	36			
3.680-90	3.686	66½	23½	3.686	59¾	30¼	3.682	54¼	35¾	3.682	49¼	40¾
3.690-00	3.699	67	23	3.695	60	30	3.693	54½	35½	3.696	49½	40½
3.700-10	3.706	67¼	22¾	3.704	60¼	29¾	3.705	54¾	35¼	3.710	49¾	40¼
3.710-20	3.712	67½	22½	3.714	60½	29½	3.716	55	35			
3.720-30	3.726	68	22	3.723	60¾	29¼	3.728	55¼	34¾	3.724	50	40
3.730-40	3.739	68½	21½	3.732	61	29	3.739	55½	34½	3.737	50¼	39¾
3.740-50	3.745	68¾	21¼	3.749	61½	28½						
3.750-60	3.752	69	21	3.759	61¾	28¼	3.750	55¾	34¼	3.751	50½	39½
3.760-70	3.764	69½	20½	3.768	62	28	3.761	56	34	3.764	50¾	39¼
3.770-80	3.777	70	20	3.776	62¼	27¾	3.772	56¼	33¾	3.778	51	39
3.780-90	3.788	70½	19½	3.785	62½	27½	3.783	56½	33½			
3.790-00	3.794	70¾	19¼	3.793	62¾	27¼	3.794	56¾	33¼	3.791	51¼	38¾
3.800-10	3.800	71	19	3.802	63	27	3.805	57	33	3.804	51½	38½
3.810-20	3.811	71½	18½	3.819	63½	26½	3.816	57¼	32¾	3.817	51¾	38¼
3.820-30	3.822	72	18	3.827	63¾	26¼	3.827	57½	32½			
3.830-40	3.833	72½	17½	3.835	64	26	3.837	57¾	32¼	3.830	52	38
3.840-50	3.843	73	17	3.843	64¼	25¾	3.847	58	32	3.843	52¼	37¾
3.850-60	3.854	73½	16½	3.851	64½	25½	3.858	58¼	31¾	3.856	52½	37½
3.860-70	3.863	74	16	3.867	65	25	3.868	58½	31½	3.869	52¾	37¼
3.870-80	3.873	74½	15½	3.875	65¼	24¾	3.878	58¾	31¼			
3.880-90	3.882	75	15	3.883	65½	24½	3.889	59	31	3.882	53	37
3.890-00	3.891	75½	14½	3.898	66	24	3.899	59¼	30¾	3.895	53¼	36¾

Approximate Lead	Gear on Worm 56 1st on Stud 48 2nd on Stud 44 Gear on Screw 100	Angle to Set Spiral Head (Degrees)	Angle to Set Vertical Att. (Degrees)	Gear on Worm 86 1st on Stud 44 2nd on Stud 28 Gear on Screw 100	Angle to Set Spiral Head (Degrees)	Angle to Set Vertical Att. (Degrees)	Gear on Worm 100 1st on Stud 48 2nd on Stud 24 Gear on Screw 86	Angle to Set Spiral Head (Degrees)	Angle to Set Vertical Att. (Degrees)	Gear on Worm 72 1st on Stud 56 2nd on Stud 48 Gear on Screw 100	Angle to Set Spiral Head (Degrees)	Angle to Set Vertical Att. (Degrees)
3.600-10				3.608	41¼	48¾				3.606	35¾	54¼
3.610-20	3.614	44¾	45¼				3.619	38½	51½			
3.620-30				3.626	41½	48½				3.627	36	54
3.630-40	3.630	45	45				3.639	38¾	51¼			
3.640-50	3.645	45¼	44¾	3.644	41¾	48¼				3.649	36¼	53¾
3.650-60							3.659	39	51			
3.660-70	3.661	45½	44½	3.662	42	48						
3.670-80	3.677	45¾	44¼				3.678	39¼	50¾	3.671	36½	53½
3.680-90				3.680	42¼	47¾						
3.690-00	3.692	46	44	3.697	42½	47½	3.698	39½	50½	3.692	36¾	53¼
3.700-10	3.707	46¼	43¾									
3.710-20				3.715	42¾	47¼	3.717	39¾	50¼	3.714	37	53
3.720-30	3.723	46½	43½									
3.730-40	3.738	46¾	43¼	3.732	43	47	3.737	40	50	3.735	37¼	52¾
3.740-50												
3.750-60	3.754	47	43	3.750	43¼	46¾	3.756	40¼	49¾	3.756	37½	52½
3.760-70	3.769	47¼	42¾	3.769	43½	46½						
3.770-80							3.776	40½	49½	3.778	37¾	52¼
3.780-90	3.784	47½	42½	3.784	43¾	46¼						
3.790-00	3.799	47¾	42¼				3.795	40¾	49¼	3.799	38	52
3.800-10				3.801	44	46						
3.810-20	3.815	48	42	3.819	44¼	45¾	3.814	41	49			
3.820-30										3.820	38¼	51¾
3.830-40	3.830	48¼	41¾	3.836	44½	45½	3.833	41¼	48¾			
3.840-50	3.845	48½	41½							3.841	38½	51½
3.850-60	3.859	48¾	41¼	3.853	44¾	45¼	3.852	41½	48½			
3.860-70										3.862	38¾	51¼
3.870-80	3.874	49	41	3.870	45	45	3.871	41¾	48¼			
3.880-90	3.888	49¼	40¾	3.887	45¼	44¾				3.883	39	51
3.890-00							3.890	42	48			

Approximate Lead	Gear on Worm 64, 1st on Stud 48, 2nd on Stud 32, Gear on Screw 100	Angle to Set Spiral Head (Degrees)	Angle to Set Vertical Att. (Degrees)	Gear on Worm 56, 1st on Stud 48, 2nd on Stud 28, Gear on Screw 72	Angle to Set Spiral Head (Degrees)	Angle to Set Vertical Att. (Degrees)	Gear on Worm 56, 1st on Stud 64, 2nd on Stud 40, Gear on Screw 72	Angle to Set Spiral Head (Degrees)	Angle to Set Vertical Att. (Degrees)	Gear on Worm 56, 1st on Stud 48, 2nd on Stud 44, Gear on Screw 100	Angle to Set Spiral Head (Degrees)	Angle to Set Vertical Att. (Degrees)	Gear on Worm 86, 1st on Stud 44, 2nd on Stud 28, Gear on Screw 100	Angle to Set Spiral Head (Degrees)	Angle to Set Vertical Att. (Degrees)	Gear on Worm 100, 1st on Stud 48, 2nd on Stud 24, Gear on Screw 86	Angle to Set Spiral Head (Degrees)	Angle to Set Vertical Att. (Degrees)	Gear on Worm 72, 1st on Stud 56, 2nd on Stud 48, Gear on Screw 100	Angle to Set Spiral Head (Degrees)	Angle to Set Vertical Att. (Degrees)	Gear on Worm 72, 1st on Stud 56, 2nd on Stud 44, Gear on Screw 86	Angle to Set Spiral Head (Degrees)	Angle to Set Vertical Att. (Degrees)
3.900-10	3.905	66¼	23¾	3.909	59½	30½	3.907	53½	36½	3.903	49½	40½	3.904	45½	44½				3.904	39¼	50¾			
3.910-20	3.913	66½	23½	3.919	59¾	30¼				3.917	49¾	40¼				3.910	42¼	47¾				3.913	36½	53½
3.920-30	3.927	67	23	3.929	60	30	3.920	53¾	36¼				3.920	45¾	44¼	3.928	42½	47½	3.925	39½	50½			
3.930-40	3.935	67¼	22¾	3.939	60¼	29¾	3.933	54	36	3.932	50	40	3.937	46	44							3.935	36¾	53¼
3.940-50	3.942	67½	22½	3.949	60½	29½	3.945	54¼	35¾	3.946	50¼	39¾				3.946	42¾	47¼	3.946	39¾	50¼			
3.950-60	3.956	68	22	3.958	60¾	29¼	3.956	54½	35½				3.953	46¼	43¾							3.959	37	53
3.960-70	3.963	68¼	21¾	3.968	61	29	3.969	54¾	35¼	3.961	50½	39½				3.965	43	47	3.966	40	50			
3.970-80	3.970	68½	21½	3.977	61¼	28¾				3.975	50¾	39¼	3.970	46½	43½									
3.980-90	3.983	69	21	3.987	61½	28½	3.982	55	35	3.989	51	39	3.986	46¾	43¼	3.984	43¼	46¾	3.987	40¼	49¾	3.981	37¼	52¾
3.990-00	3.996	69½	20½	3.997	61¾	28¼	3.994	55¼	34¾															
4.000-10	4.003	69¾	20¼	4.006	62	28	4.006	55½	34½	4.003	51¼	38¾	4.003	47	43	4.002	43½	46½	4.008	40½	49½	4.004	37½	52½
4.010-20	4.010	70	20	4.015	62¼	27¾	4.018	55¾	34¼	4.017	51½	38½	4.019	47¼	42¾									
4.020-30	4.022	70½	19½	4.024	62½	27½										4.020	43¾	46¼	4.028	40¾	49¼	4.027	37¾	52¼
4.030-40	4.034	71	19	4.033	62¾	27¼	4.030	56	34	4.031	51¾	38¼	4.035	47½	42½	4.038	44	46						
4.040-50	4.046	71½	18½	4.042	63	27	4.041	56¼	33¾	4.045	52	38							4.048	41	49	4.050	38	52
4.050-60	4.058	72	18	4.052	63¼	26¾	4.054	56½	33½	4.059	52¼	37¾	4.051	47¾	42¼	4.057	44¼	45¾						
4.060-70	4.069	72½	17½	4.060	63½	26½	4.065	56¾	33¼				4.067	48	42				4.069	41¼	48¾			
4.070-80	4.075	72¾	17¼	4.078	64	26	4.077	57	33	4.072	52½	37½				4.075	44½	45½				4.072	38¼	51¾
4.080-90	4.080	73	17	4.086	64¼	25¾	4.088	57¼	32¾	4.086	52¾	37¼	4.084	48¼	41¾				4.089	41½	48½			
4.090-00	4.091	73½	16½	4.095	64½	25½							4.099	48½	41½	4.093	44¾	45¼				4.095	38½	51½
4.100-10	4.101	74	16	4.104	64¾	25¼	4.100	57½	32½	4.100	53	37							4.109	41¾	48¼			
4.110-20	4.112	74½	15½	4.112	65	25	4.111	57¾	32¼	4.113	53¼	36¾	4.115	48¾	41¼	4.111	45	45				4.117	38¾	51¼
4.120-30	4.121	75	15	4.128	65½	24½	4.122	58	32	4.126	53½	36½				4.129	45¼	44¾	4.129	42	48			
4.130-40	4.131	75½	14½	4.137	65¾	24¼	4.133	58¼	31¾				4.130	49	41									
4.140-50	4.149	76½	13½	4.145	66	24	4.144	58½	31½	4.140	53¾	36¼	4.146	49¼	40¾	4.147	45½	44½	4.149	42¼	47¾	4.140	39	51
4.150-60	4.157	77	13	4.153	66¼	23¾	4.156	58¾	31¼	4.153	54	36												
4.160-70	4.166	77½	12½	4.161	66½	23½	4.167	59	31	4.165	54¼	35¾	4.161	49½	40½	4.164	45¾	44¼	4.169	42½	47½	4.162	39¼	50¾
4.170-80	4.174	78	12	4.176	67	23	4.177	59¼	30¾	4.178	54½	35½	4.177	49¾	40¼									
4.180-90	4.181	78½	11½	4.184	67¼	22¾	4.189	59½	30½							4.182	46	44	4.189	42¾	47¼	4.184	39½	50½
4.190-00	4.195	79½	10½	4.199	67¾	22¼	4.199	59¾	30¼	4.192	54¾	35¼	4.192	50	40	4.192	46¼	43¾						

LEADS FROM 4.200 TO 4.500

Approximate Lead	Gear on Worm 56, 1st on Stud 48, 2nd on Stud 28, Gear on Screw 72	Angle to Set Spiral Head (Degrees)	Angle to Set Vertical Att. (Degrees)	Gear on Worm 56, 1st on Stud 64, 2nd on Stud 40, Gear on Screw 72	Angle to Set Spiral Head (Degrees)	Angle to Set Vertical Att. (Degrees)	Gear on Worm 56, 1st on Stud 48, 2nd on Stud 44, Gear on Screw 100	Angle to Set Spiral Head (Degrees)	Angle to Set Vertical Att. (Degrees)	Gear on Worm 86, 1st on Stud 44, 2nd on Stud 28, Gear on Screw 100	Angle to Set Spiral Head (Degrees)	Angle to Set Vertical Att. (Degrees)	Gear on Worm 100, 1st on Stud 48, 2nd on Stud 24, Gear on Screw 86	Angle to Set Spiral Head (Degrees)	Angle to Set Vertical Att. (Degrees)	Gear on Worm 72, 1st on Stud 56, 2nd on Stud 48, Gear on Screw 100	Angle to Set Spiral Head (Degrees)	Angle to Set Vertical Att. (Degrees)	Gear on Worm 72, 1st on Stud 56, 2nd on Stud 44, Gear on Screw 86	Angle to Set Spiral Head (Degrees)	Angle to Set Vertical Att. (Degrees)	Gear on Worm 56, 1st on Stud 32, 2nd on Stud 40, Gear on Screw 100	Angle to Set Spiral Head (Degrees)	Angle to Set Vertical Att. (Degrees)
4.200-10	4.206	68	22	4.209	60	30	4.204	55	35	4.208	50¼	39¾				4.208	43	47	4.206	39¾				
4.210-20	4.214	68¼	21¾				4.217	55¼	34¾				4.217	46½	43½							4.213	37	53
4.220-30	4.221	68½	21½	4.220	60¼	29¾				4.223	50½	39½				4.229	43¼	46¾	4.228	40	50			
4.230-40	4.235	69	21	4.231	60½	29½	4.230	55½	34½	4.238	50¾	39¼	4.234	46¾	43¼							4.237	37¼	52¾
4.240-50	4.249	69½	20½	4.242	60¾	29¼	4.243	55¾	34¼							4.247	43½	46½						
4.250-60	4.256	69¾	20¼	4.252	61	29	4.256	56	34	4.253	51	39	4.252	47	43				4.250	40¼	49¾			
4.260-70	4.267	70¼	19¾	4.262	61¼	28¾	4.268	56¼	33¾	4.268	51¼	38¾	4.269	47¼	42¾	4.267	43¾	46¼				4.261	37½	52½
4.270-80	4.277	70½	19½	4.272	61½	28½													4.272	40½	49½			
4.280-90	4.283	70¾	19¼	4.282	61¾	28¼	4.281	56½	33½	4.283	51½	38½	4.287	47½	42½	4.286	44	46				4.286	37¾	52¼
4.290-00	4.290	71	19	4.292	62	28	4.293	56¾	33¼	4.298	51¾	38¼							4.294	40¾	49¼			
4.300-10	4.309	71¾	18¼	4.302	62¼	27¾	4.305	57	33				4.304	47¾	42¼	4.306	44¼	45¾						
4.310-20	4.315	72	18	4.312	62½	27½	4.317	57¼	32¾	4.313	52	38							4.316	41	49	4.310	38	52
4.320-30	4.327	72½	17½	4.321	62¾	27¼	4.329	57½	32½	4.327	52¼	37¾	4.321	48	42	4.325	44½	45½						
4.330-40	4.333	72¾	17¼	4.331	63	27							4.338	48¼	41¾				4.337	41¼	48¾	4.334	38¼	51¾
4.340-50	4.344	73¼	16¾	4.341	63¼	26¾	4.341	57¾	32¼	4.342	52½	37½				4.344	44¾	45¼						
4.350-60	4.350	73½	16½	4.350	63½	26½	4.353	58	32	4.357	52¾	37¼	4.355	48½	41½				4.358	41½	48½	4.357	38½	51½
4.360-70	4.361	74	16	4.369	64	26	4.365	58¼	31¾							4.364	45	45						
4.370-80	4.377	74¾	15¼	4.378	64¼	25¾	4.376	58½	31½	4.372	53	37	4.371	48¾	41¼									
4.380-90	4.381	75	15	4.387	64½	25½	4.388	58¾	31¼	4.385	53¼	36¾	4.388	49	41	4.382	45¼	44¾	4.380	41¾	48¼	4.381	38¾	51¼
4.390-00	4.392	75½	14½	4.396	64¾	25¼				4.399	53½	36½												
4.400-10	4.407	76¼	13¾	4.405	65	25	4.400	59	31				4.404	49¼	40¾	4.401	45½	44½	4.402	42	48	4.405	39	51
4.410-20	4.416	76¾	13¼	4.414	65¼	24¾	4.411	59¼	30¾	4.413	53¾	36¼												
4.420-30	4.424	77¼	12¾	4.423	65½	24½	4.422	59½	30½	4.428	54	36	4.421	49½	40½	4.420	45¾	44¼	4.423	42¼	47¾	4.429	39¼	50¾
4.430-40	4.438	78	12	4.432	65¾	24¼	4.434	59¾	30¼				4.437	49¾	40¼	4.439	46	44						
4.440-50	4.442	78½	11½	4.440	66	24	4.445	60	30	4.442	54¼	35¾							4.444	42½	47½			
4.450-60	4.453	79	11	4.457	66½	23½	4.456	60¼	29¾	4.455	54½	35½	4.454	50	40	4.457	46¼	43¾				4.453	39½	50½
4.460-70	4.461	79½	10½	4.466	66¾	23¼	4.468	60½	29½	4.469	54¾	35¼							4.465	42¾	47¼			
4.470-80				4.474	67	23	4.478	60¾	29¼				4.470	50¼	39¾	4.476	46½	43½				4.476	39¾	50¼
4.480-90				4.483	67¼	22¾	4.489	61	29	4.483	55	35	4.486	50½	39½				4.486	43	47			
4.490-00				4.491	67½	22½				4.497	55¼	34¾				4.495	46¾	43¼				4.499	40	50

LEADS FROM 4.500 TO 4.800

Approximate Lead	Gear on Worm 56, 1st on Stud 64, 2nd on Stud 40, Gear on Screw 72	Angle to Set Spiral Head (Degrees)	Angle to Set Vertical Att. (Degrees)	Gear on Worm 56, 1st on Stud 48, 2nd on Stud 44, Gear on Screw 100	Angle to Set Spiral Head (Degrees)	Angle to Set Vertical Att. (Degrees)	Gear on Worm 86, 1st on Stud 44, 2nd on Stud 28, Gear on Screw 100	Angle to Set Spiral Head (Degrees)	Angle to Set Vertical Att. (Degrees)	Gear on Worm 100, 1st on Stud 48, 2nd on Stud 24, Gear on Screw 86	Angle to Set Spiral Head (Degrees)	Angle to Set Vertical Att. (Degrees)	Gear on Worm 72, 1st on Stud 56, 2nd on Stud 48, Gear on Screw 100	Angle to Set Spiral Head (Degrees)	Angle to Set Vertical Att. (Degrees)	Gear on Worm 72, 1st on Stud 56, 2nd on Stud 44, Gear on Screw 86	Angle to Set Spiral Head (Degrees)	Angle to Set Vertical Att. (Degrees)	Gear on Worm 56, 1st on Stud 32, 2nd on Stud 40, Gear on Screw 100	Angle to Set Spiral Head (Degrees)	Angle to Set Vertical Att. (Degrees)	Gear on Worm 56, 1st on Stud 28, 2nd on Stud 32, Gear on Screw 86	Angle to Set Spiral Head (Degrees)	Angle to Set Vertical Att. (Degrees)
4.500-10	4.507	68	22	4.500	61¼	28¾				4.502	50¾	39¼				4.507	43¼	46¾				4.504	37¼	52¾
4.510-20	4.515	68¼	21¾	4.511	61½	28½	4.510	55½	34½	4.518	51	39	4.513	47	43									
4.520-30	4.523	68½	21½	4.521	61¾	28¼	4.524	55¾	34¼							4.528	43½	46½	4.523	40¼	49¾			
4.530-40	4.538	69	21	4.532	62	28	4.537	56	34	4.534	51¼	38¾	4.531	47¼	42¾							4.530	37½	52½
4.540-50	4.545	69¼	20¾	4.543	62¼	27¾										4.549	43¾	46¼	4.546	40½	49½			
4.550-60	4.553	69½	20½	4.553	62½	27½	4.550	56¼	33¾	4.550	51½	38½	4.550	47½	42½							4.556	37¾	52¼
4.560-70	4.561	69¾	20¼	4.565	62¾	27¼	4.564	56½	33½	4.566	51¾	38¼	4.568	47¾	42¼	4.569	44	46	4.569	40¾	49¼			
4.570-80	4.575	70¼	19¾	4.573	63	27	4.577	56¾	33¼															
4.580-90	4.582	70½	19½	4.584	63¼	26¾				4.581	52	38	4.586	48	42							4.582	38	52
4.590-00	4.596	71	19	4.593	63½	26½	4.590	57	33	4.597	52¼	37¾				4.590	44¼	45¾	4.592	41	49			
4.600-10	4.610	71¼	18¾	4.603	63¾	26¼	4.603	57¼	32¾				4.604	48¼	41¾							4.607	38¼	51¾
4.610-20	4.616	71¾	18¼	4.613	64	26	4.616	57½	32½	4.612	52½	37½				4.610	44½	45½	4.615	41¼	48¾			
4.620-30	4.629	72¼	17¾	4.623	64¼	25¾	4.629	57¾	32¼	4.628	52¾	37¼	4.622	48½	41½									
4.630-40	4.636	72½	17½	4.633	64½	25½							4.639	48¾	41¼	4.631	44¾	45¼	4.638	41½	48½	4.633	38½	51½
4.640-50	4.648	73	17	4.642	64¾	25¼	4.641	58	32	4.643	53	37												
4.650-60	4.654	73¼	16¾	4.652	65	25	4.654	58¼	31¾	4.658	53¼	36¾	4.657	49	41	4.651	45	45				4.658	38¾	51¼
4.660-70	4.660	73½	16½	4.661	65¼	24¾	4.666	58½	31½										4.661	41¾	48¼			
4.670-80	4.672	74	16	4.671	65½	24½	4.678	58¾	31¼	4.673	53½	36½	4.674	49¼	40¾	4.672	45¼	44¾						
4.680-90	4.689	74¾	15¼	4.680	65¾	24¼				4.689	53¾	36¼							4.684	42	48	4.683	39	51
4.690-00	4.695	75	15	4.698	66¼	23¾	4.691	59	31				4.692	49½	40½	4.692	45½	44½						
4.700-10	4.701	75¼	15¾	4.707	66½	23½	4.703	59¼	30¾	4.704	54	36	4.709	49¾	40¼				4.707	42¼	47¾	4.708	39¼	50¾
4.710-20	4.711	75¾	14¼	4.716	66¾	23¼	4.715	59½	30½	4.718	54¼	35¾				4.712	45¾	44¼						
4.720-30	4.720	76¼	13¾	4.725	67	23	4.727	59¾	30¼				4.727	50	40				4.729	42½	47½			
4.730-40	4.736	77	13	4.733	67¼	22¾				4.733	54½	35½				4.731	46	44				4.734	39½	50½
4.740-50	4.746	77½	12½	4.742	67½	22½	4.740	60	30	4.748	54¾	35¼	4.744	50¼	39¾									
4.750-60	4.750	77¾	12¼	4.759	68	22	4.752	60¼	29¾							4.751	46¼	43¾	4.752	42¾	47¼	4.758	39¾	50¼
4.760-70	4.763	78½	11½	4.767	68¼	21¾	4.764	60½	29½	4.762	55	35	4.762	50½	39½									
4.770-80	4.772	79	11	4.774	68½	21½	4.775	60¾	29¼	4.777	55¼	34¾	4.779	50¾	39¼	4.771	46½	43½	4.774	43	47			
4.780-90	4.783	79¾	10¼	4.784	68¾	21¼	4.787	61	29															
4.790-00				4.793	69	21	4.798	61¼	28¾	4.792	55½	34½	4.796	51	39	4.791	46¾	43¼	4.796	43¼	46¾			

LEADS FROM 4.800 TO 5.100

Approximate Lead	Gear on Worm 56, 1st on Stud 48, 2nd on Stud 44, Gear on Screw 100	Angle to Set Spiral Head (Degrees)	Angle to Set Vertical Att. (Degrees)	Gear on Worm 86, 1st on Stud 44, 2nd on Stud 28, Gear on Screw 100	Angle to Set Spiral Head (Degrees)	Angle to Set Vertical Att. (Degrees)	Gear on Worm 100, 1st on Stud 48, 2nd on Stud 24, Gear on Screw 86	Angle to Set Spiral Head (Degrees)	Angle to Set Vertical Att. (Degrees)	Gear on Worm 72, 1st on Stud 56, 2nd on Stud 48, Gear on Screw 100	Angle to Set Spiral Head (Degrees)	Angle to Set Vertical Att. (Degrees)	Gear on Worm 72, 1st on Stud 56, 2nd on Stud 44, Gear on Screw 86	Angle to Set Spiral Head (Degrees)	Angle to Set Vertical Att. (Degrees)	Gear on Worm 56, 1st on Stud 32, 2nd on Stud 40, Gear on Screw 100	Angle to Set Spiral Head (Degrees)	Angle to Set Vertical Att. (Degrees)	Gear on Worm 56, 1st on Stud 28, 2nd on Stud 32, Gear on Screw 86	Angle to Set Spiral Head (Degrees)	Angle to Set Vertical Att. (Degrees)	Gear on Worm 72, 1st on Stud 40, 2nd on Stud 44, Gear on Screw 100	Angle to Set Spiral Head (Degrees)	Angle to Set Vertical Att. (Degrees)
4.800-10	4.808	69½	20½				4.806	55¾	34¼										4.808	40¼	49¾			
4.810-20	4.816	69¾	20¼	4.810	61½	28½				4.813	51¼	38¾	4.812	47	43	4.818	43½	46½						
4.820-30	4.824	70	20	4.821	61¾	28¼	4.820	56	34	4.829	51½	38½										4.821	37½	52½
4.830-40	4.839	70½	19½	4.833	62	28	4.834	56¼	33¾				4.830	47¼	42¾				4.833	40½	49½			
4.840-50	4.846	70¾	19¼	4.843	62¼	27¾	4.848	56½	33½	4.846	51¾	38¼				4.841	43¾	46¼				4.849	37¾	52¼
4.850-60	4.853	71	19	4.854	62½	27½							4.850	47½	42½				4.858	40¾	49¼			
4.860-70	4.865	71¼	18¾	4.865	62¾	27¼	4.862	56¾	33¼	4.863	52	38	4.869	47¾	42¼	4.862	44	46						
4.870-80	4.875	71¾	18¼	4.876	63	27	4.876	57	33	4.879	52¼	37¾										4.876	38	52
4.880-90	4.881	72	18	4.887	63¼	26¾							4.888	48	42	4.885	44¼	45¾	4.882	41	49			
4.890-00	4.895	72½	17½	4.898	63½	26½	4.890	57¼	32¾	4.896	52½	37½												
4.900-10	4.902	72¾	17¼	4.908	63¾	26¼	4.903	57½	32½				4.908	48¼	41¾	4.906	44½	45½	4.907	41¼	48¾	4.903	38¼	51¾
4.910-20	4.915	73¼	17¾	4.919	64	26	4.917	57¾	32¼	4.912	52¾	37¼												
4.920-30	4.921	73½	17½	4.929	64¼	25¾				4.928	53	37	4.927	48½	41½	4.928	44¾	45¼						
4.930-40	4.934	74	16				4.930	58	32										4.931	41½	48½	4.930	38½	51½
4.940-50	4.946	74½	15½	4.940	64½	25½	4.944	58¼	31¾	4.944	53¼	36¾	4.945	48¾	41¼									
4.950-60	4.952	74¾	15¼	4.950	64¾	25¼	4.957	58½	31½							4.950	45	45	4.955	41¾	48¼	4.957	38¾	51¼
4.960-70	4.964	75¼	14¾	4.960	65	25				4.961	53½	36½	4.964	49	41									
4.970-80	4.975	75¾	14¼	4.970	65¼	24¾	4.970	58¾	31¼	4.976	53¾	36¼				4.971	45¼	44¾						
4.980-90	4.986	76¼	13¾	4.980	65½	24½	4.984	59	31				4.983	49¼	40¾				4.980	42	48	4.984	39	51
4.990-00	4.996	76½	13½	4.990	65¾	24¼	4.996	59¼	30¾	4.992	54	36				4.993	45½	44½						
5.000-10	5.001	77	13	5.000	66	24	5.009	59½	30½	5.008	54¼	35¾	5.002	49½	40½				5.004	42¼	47¾			
5.010-20	5.016	77¾	12¼	5.019	66½	23½										5.014	45¾	44¼				5.011	39¼	50¾
5.020-30	5.025	78¼	11¾	5.028	66¾	23¼	5.022	59¾	30¼	5.024	54½	35½	5.020	49¾	40¼				5.028	42½	47½			
5.030-40	5.030	78½	11½	5.038	67	23	5.035	60	30	5.039	54¾	35¼	5.039	50	40	5.035	46	44				5.038	39½	50½
5.040-50	5.043	79¼	10¾	5.047	67¼	22¾	5.048	60¼	29¾															
5.050-60	5.051	79¾	10¼	5.056	67½	22½				5.055	55	35	5.058	50¼	39¾	5.056	46¼	43¾	5.052	42¾	47¼			
5.060-70				5.065	67¾	22¼	5.061	60½	29½													5.064	39¾	50¼
5.070-80				5.072	68	22	5.073	60¾	29¼	5.070	55¼	34¾	5.076	50½	39½	5.077	46½	43½	5.075	43	47			
5.080-90				5.083	68¼	21¾	5.085	61	29	5.086	55½	34½												
5.090-00				5.092	68½	21½	5.097	61¼	28¾				5.094	50¾	39¼	5.098	46¾	43¼	5.099	43¼	46¾	5.091	40	50

LEADS FROM 5.100 TO 5.400

Approximate Lead	Lead	Angle to Set Spiral Head	Angle to Set Vertical Att.	Lead	Angle to Set Spiral Head	Angle to Set Vertical Att.	Lead	Angle to Set Spiral Head	Angle to Set Vertical Att.	Lead	Angle to Set Spiral Head	Angle to Set Vertical Att.
Gear on Worm / 1st on Stud / 2nd on Stud / Gear on Screw	86 / 44 / 28 / 100	Degrees		100 / 48 / 24 / 86	Degrees		72 / 56 / 48 / 100	Degrees		72 / 56 / 44 / 86	Degrees	
5.100-10	5.109	69	21				5.101	55¾	34¼			
5.110-20	5.118	69¼	20¾	5.110	61½	28½	5.117	56	34	5.112	51	39
5.120-30	5.126	69½	20½	5.121	61¾	28¼						
5.130-40	5.135	69¾	20¼	5.133	62	28	5.131	56¼	33¾	5.130	51¼	38¾
5.140-50	5.143	70	20	5.145	62¼	27¾	5.146	56½	33½	5.148	51½	38½
5.150-60	5.151	70¼	19¾	5.157	62½	27½						
5.160-70	5.167	70¾	19¼	5.169	62¾	27¼	5.161	56¾	33¼	5.166	51¾	38¼
5.170-80	5.175	71	19				5.176	57	33			
5.180-90	5.182	71¼	18¾	5.180	63	27				5.183	52	38
5.190-00	5.190	71½	18½	5.192	63¼	26¾	5.191	57¼	32¾			
5.200-10	5.204	72	18	5.203	63½	26½	5.204	57½	32½	5.205	52¼	37¾
5.210-20	5.212	72¼	17¾	5.214	63¾	26¼	5.219	57¾	32¼	5.218	52½	37½
5.220-30	5.226	72¾	17¼	5.225	64	26						
5.230-40	5.233	73	17	5.236	64¼	25¾	5.233	58	32	5.236	52¾	37¼
5.240-50	5.240	73¼	16¾	5.247	64½	25½	5.247	58¼	31¾			
5.250-60	5.254	73¾	16¼	5.258	64¾	25¼				5.253	53	37
5.260-70	5.261	74	16	5.269	65	25	5.261	58½	31½			
5.270-80	5.274	74½	15½				5.275	58¾	31¼	5.270	53¼	36¾
5.280-90	5.286	75	15	5.280	65¼	24¾				5.288	53½	36½
5.290-00	5.298	75½	14½	5.290	65½	24½	5.290	59	31			
5.300-10	5.304	75¾	14¼	5.300	65¾	24¼	5.303	59¼	30¾	5.305	53¾	36¼
5.310-20	5.310	76	14	5.311	66	24	5.316	59½	30½			
5.320-30	5.327	76¾	13¼	5.321	66¼	23¾				5.322	54	36
5.330-40	5.332	77	13	5.332	66½	23½	5.330	59¾	30¼	5.338	54¼	35¾
5.340-50	5.348	77¾	12¼	5.342	66¾	23¼	5.344	60	30			
5.350-60	5.353	78	12	5.352	67	23	5.358	60¼	29¾	5.355	54½	35½
5.360-70	5.368	78¾	11¼	5.362	67¼	22¾						
5.370-80	5.377	79¼	10¾	5.372	67½	22½	5.371	60½	29½	5.372	54¾	35¼
5.380-90	5.385	79¾	10¼	5.381	67¾	22¼	5.384	60¾	29¼	5.388	55	35
5.390-00	5.390	80	10	5.391	68	22	5.397	61	29			

Approximate Lead	Lead	Angle to Set Spiral Head	Angle to Set Vertical Att.	Lead	Angle to Set Spiral Head	Angle to Set Vertical Att.	Lead	Angle to Set Spiral Head	Angle to Set Vertical Att.	Lead	Angle to Set Spiral Head	Angle to Set Vertical Att.
Gear on Worm / 1st on Stud / 2nd on Stud / Gear on Screw	56 / 32 / 40 / 100	Degrees		56 / 28 / 32 / 86	Degrees		72 / 40 / 44 / 100	Degrees		56 / 32 / 48 / 100	Degrees	
5.100-10												
5.110-20							5.117	40¼	49¾	5.114	37½	52½
5.120-30	5.120	47	43	5.122	43½	46½						
5.130-40												
5.140-50	5.140	47¼	42¾	5.146	43¾	46¼	5.144	40½	49½	5.143	37¾	52¼
5.150-60												
5.160-70	5.161	47½	42½	5.169	44	46						
5.170-80							5.170	40¾	49¼	5.172	38	52
5.180-90	5.182	47¾	42¼									
5.190-00				5.193	44¼	45¾	5.196	41	49			
5.200-10	5.202	48	42							5.200	38¼	51¾
5.210-20				5.216	44½	45½						
5.220-30	5.223	48¼	41¾				5.222	41¼	48¾	5.229	38½	51½
5.230-40				5.239	44¾	45¼						
5.240-50	5.243	48½	41½				5.248	41½	48½			
5.250-60										5.258	38¾	51¼
5.260-70	5.263	48¾	41¼	5.262	45	45						
5.270-80							5.273	41¾	48¼			
5.280-90	5.283	49	41	5.285	45¼	44¾				5.286	39	51
5.290-00												
5.300-10	5.302	49¼	40¾	5.308	45½	44½	5.300	42	48			
5.310-20										5.315	39¼	50¾
5.320-30	5.323	49½	40½				5.326	42¼	47¾			
5.330-40				5.331	45¾	44¼						
5.340-50	5.342	49¾	40¼							5.343	39½	50½
5.350-60				5.353	46	44	5.350	42½	47½			
5.360-70	5.362	50	40									
5.370-80				5.375	46¼	43¾	5.376	42¾	47¼	5.371	39¾	50¼
5.380-90	5.382	50¼	39¾									
5.390-00				5.398	46½	43½				5.399	40	50

LEADS FROM 5.400 TO 5.700

Approximate Lead	Gear on Worm 100, 1st on Stud 48, 2nd on Stud 24, Gear on Screw 86	Angle to Set Spiral Head (Degrees)	Angle to Set Vertical Att. (Degrees)	Gear on Worm 72, 1st on Stud 56, 2nd on Stud 48, Gear on Screw 100	Angle to Set Spiral Head (Degrees)	Angle to Set Vertical Att. (Degrees)	Gear on Worm 72, 1st on Stud 56, 2nd on Stud 44, Gear on Screw 86	Angle to Set Spiral Head (Degrees)	Angle to Set Vertical Att. (Degrees)	Gear on Worm 56, 1st on Stud 32, 2nd on Stud 40, Gear on Screw 100	Angle to Set Spiral Head (Degrees)	Angle to Set Vertical Att. (Degrees)
5.400-10	5.400	68¼	21¾				5.405	55¼	34¾	5.401	50½	39½
5.410-20	5.419	68¾	21¼	5.410	61¼	28¾						
5.420-30	5.428	69	21	5.423	61½	28½	5.421	55½	34½	5.421	50¾	39¼
5.430-40	5.437	69¼	20¾	5.436	61¾	28¼	5.437	55¾	34¼			
5.440-50	5.446	69½	20½	5.448	62	28				5.440	51	39
5.450-60	5.455	69¾	20¼				5.454	56	34	5.459	51¼	38¾
5.460-70	5.464	70	20	5.461	62¼	27¾	5.469	56¼	33¾			
5.470-80	5.472	70¼	19¾	5.473	62½	27½				5.478	51½	38½
5.480-90	5.489	70¾	19¼	5.486	62¾	27¼	5.486	56½	33½			
5.490-00	5.497	71	19	5.498	63	27				5.497	51¾	38¼
5.500-10	5.505	71¼	18¾				5.501	56¾	33¼			
5.510-20	5.513	71½	18½	5.510	63¼	26¾	5.517	57	33	5.516	52	38
5.520-30	5.521	71¾	18¼	5.522	63½	26½						
5.530-40	5.537	72¼	17¾	5.534	63¾	26¼	5.532	57¼	32¾	5.535	52¼	37¾
5.540-50	5.545	72½	17½	5.546	64	26	5.548	57½	32½			
5.550-60	5.552	72¾	17¼	5.558	64¼	25¾				5.553	52½	37½
5.560-70	5.560	73	17				5.563	57¾	32¼			
5.570-80	5.574	73½	16½	5.570	64½	25½	5.578	58	32	5.572	52¾	37¼
5.580-90	5.588	74	16	5.581	64¾	25¼						
5.590-00	5.595	74¼	15¾	5.592	65	25	5.594	58¼	31¾	5.590	53	37
5.600-10	5.602	74½	15½	5.604	65¼	24¾	5.608	58½	31½	5.609	53¼	36¾
5.610-20	5.616	75	15	5.615	65½	24½						
5.620-30	5.629	75½	14½	5.626	65¾	24¼	5.623	58¾	31¼	5.627	53½	36½
5.630-40	5.635	75¾	14¼	5.637	66	24	5.638	59	31			
5.640-50	5.641	76	14	5.648	66¼	23¾				5.645	53¾	36¼
5.650-60	5.659	76¾	13¼	5.658	66½	23½	5.653	59¼	30¾			
5.660-70	5.662	77	13				5.667	59½	30½	5.663	54	36
5.670-80	5.676	77½	12½	5.670	66¾	23¼						
5.680-90	5.687	78	12	5.680	67	23	5.682	59¾	30¼	5.681	54¼	35¾
5.690-00	5.692	78¼	11¾	5.691	67¼	22¾	5.698	60	30	5.699	54½	35½

Approximate Lead	Gear on Worm 56, 1st on Stud 28, 2nd on Stud 32, Gear on Screw 86	Angle to Set Spiral Head (Degrees)	Angle to Set Vertical Att. (Degrees)	Gear on Worm 72, 1st on Stud 40, 2nd on Stud 44, Gear on Screw 100	Angle to Set Spiral Head (Degrees)	Angle to Set Vertical Att. (Degrees)	Gear on Worm 56, 1st on Stud 32, 2nd on Stud 48, Gear on Screw 100	Angle to Set Spiral Head (Degrees)	Angle to Set Vertical Att. (Degrees)	Gear on Worm 86, 1st on Stud 32, 2nd on Stud 24, Gear on Screw 72	Angle to Set Spiral Head (Degrees)	Angle to Set Vertical Att. (Degrees)
5.400-10				5.401	43	47						
5.410-20												
5.420-30	5.420	46¾	43¼	5.427	43¼	46¾	5.427	40¼	49¾	5.423	37¼	52¾
5.430-40												
5.440-50	5.443	47	43									
5.450-60				5.451	43½	46½	5.455	40½	49½	5.454	37½	52½
5.460-70	5.465	47¼	42¾									
5.470-80				5.477	43¾	46¼						
5.480-90	5.487	47½	42½				5.483	40¾	49¼	5.485	37¾	52¼
5.490-00												
5.500-10	5.509	47¾	42¼	5.501	44	46						
5.510-20							5.511	41	49	5.516	38	52
5.520-30				5.527	44¼	45¾						
5.530-40	5.530	48	42				5.538	41¼	48¾			
5.540-50										5.547	38¼	51¾
5.550-60	5.552	48¼	41¾	5.550	44½	45½						
5.560-70							5.565	41½	48½			
5.570-80	5.574	48½	41½	5.576	44¾	45¼				5.577	38½	51½
5.580-90												
5.590-00	5.595	48¾	41¼				5.593	41¾	48¼			
5.600-10				5.600	45	45				5.607	38¾	51¼
5.610-20	5.616	49	41									
5.620-30				5.624	45¼	44¾	5.621	42	48			
5.630-40	5.637	49¼	40¾							5.638	39	51
5.640-50				5.649	45½	44½	5.648	42¼	47¾			
5.650-60	5.659	49½	40½									
5.660-70										5.668	39¼	50¾
5.670-80				5.673	45¾	44¼	5.675	42½	47½			
5.680-90	5.680	49¾	40¼									
5.690-00				5.697	46	44				5.699	39½	50½

Approximate Lead	Gear on Worm 72, 1st on Stud 56, 2nd on Stud 48, Gear on Screw 100	Angle to Set Spiral Head (Degrees)	Angle to Set Vertical Att. (Degrees)	Gear on Worm 72, 1st on Stud 56, 2nd on Stud 44, Gear on Screw 86	Angle to Set Spiral Head (Degrees)	Angle to Set Vertical Att. (Degrees)	Gear on Worm 56, 1st on Stud 32, 2nd on Stud 40, Gear on Screw 100	Angle to Set Spiral Head (Degrees)	Angle to Set Vertical Att. (Degrees)	Gear on Worm 56, 1st on Stud 28, 2nd on Stud 32, Gear on Screw 86	Angle to Set Spiral Head (Degrees)	Angle to Set Vertical Att. (Degrees)	Gear on Worm 72, 1st on Stud 40, 2nd on Stud 44, Gear on Screw 100	Angle to Set Spiral Head (Degrees)	Angle to Set Vertical Att. (Degrees)	Gear on Worm 56, 1st on Stud 32, 2nd on Stud 48, Gear on Screw 100	Angle to Set Spiral Head (Degrees)	Angle to Set Vertical Att. (Degrees)	Gear on Worm 86, 1st on Stud 32, 2nd on Stud 24, Gear on Screw 72	Angle to Set Spiral Head (Degrees)	Angle to Set Vertical Att. (Degrees)	Gear on Worm 64, 1st on Stud 48, 2nd on Stud 40, Gear on Screw 56	Angle to Set Spiral Head (Degrees)	Angle to Set Vertical Att. (Degrees)
5.700-10	5.701	67½	22½							5.701	50	40				5.702	42¾	47¼						
5.710-20	5.711	67¾	22¼	5.711	60¼	29¾	5.716	54¾	35¼															
5.720-30	5.722	68	22	5.726	60½	29½				5.722	50¼	39¾	5.721	46¼	43¾	5.729	43	47	5.729	39¾	50¼			
5.730-40	5.731	68¼	21¾	5.739	60¾	29¼	5.734	55	35													5.732	37	53
5.740-50	5.741	68½	21½							5.742	50½	39½	5.745	46½	43½									
5.750-60	5.751	68¾	21¼	5.753	61	29	5.759	55¼	34¾							5.756	43¼	46¾	5.759	40	50			
5.760-70	5.761	69	21	5.767	61¼	28¾	5.769	55½	34½	5.763	50¾	39¼	5.766	46¾	43¼							5.765	37¼	52¾
5.770-80	5.771	69¼	20¾																					
5.780-90	5.780	69½	20½	5.781	61½	28½	5.786	55¾	34¼	5.784	51	39				5.782	43½	46½	5.788	40¼	49¾			
5.790-00	5.799	70	20	5.792	61¾	28¼							5.792	47	43							5.798	37½	52½
5.800-10	5.808	70¼	19¾	5.808	62	28	5.804	56	34	5.804	51¼	38¾				5.808	43¾	46¼						
5.810-20	5.817	70½	19½										5.816	47¼	42¾				5.818	40½	49½			
5.820-30	5.826	70¾	19¼	5.822	62¼	27¾	5.820	56¼	33¾	5.824	51½	38½												
5.830-40	5.835	71	19	5.834	62½	27½	5.837	56½	33½							5.835	44	46				5.8307	37¾	52¼
5.840-50	5.843	71¼	18¾	5.848	62¾	27¼				5.844	51¾	38¼	5.840	47½	42½				5.848	40¾	49¼			
5.850-60	5.852	71½	18½				5.854	56¾	33¼															
5.860-70	5.860	71¾	18¼	5.861	63	27				5.864	52	38	5.863	47¾	42¼	5.862	44¼	45¾				5.863	38	52
5.870-80	5.877	72¼	17¾	5.874	63¼	26¾	5.871	57	33										5.877	41	49			
5.880-90	5.885	72½	17½	5.886	63½	26½	5.887	57¼	32¾	5.884	52¼	37¾	5.886	48	42	5.888	44½	45½						
5.890-00	5.893	72¾	17¼	5.899	63¾	26¼																5.896	38¼	51¾
5.900-10	5.901	73	17				5.904	57½	32½	5.904	52½	37½	5.909	48¼	41¾				5.907	41¼	48¾			
5.910-20	5.916	73½	16½	5.912	64	26										5.914	44¾	45¼						
5.920-30	5.924	73¾	16¼	5.924	64¼	25¾	5.920	57¾	32¼	5.924	52¾	37¼										5.928	38½	51½
5.930-40	5.932	74	16	5.937	64½	25½	5.936	58	32				5.932	48½	41½				5.936	41½	48½			
5.940-50	5.946	74½	15½	5.949	64¾	25¼				5.944	53	37				5.940	45	45						
5.950-60	5.954	74¾	15¼				5.952	58¼	31¾				5.954	48¾	41¼									
5.960-70	5.967	75¼	14¾	5.961	65	25	5.968	58½	31½	5.963	53¼	36¾				5.966	45¼	44¾	5.965	41¾	48¼	5.960	38¾	51¼
5.970-80	5.974	75½	14½	5.974	65¼	24¾							5.977	49	41									
5.980-90	5.981	75¾	14¼	5.986	65½	24½	5.984	58¾	31¼	5.982	53½	36½												
5.990-00	5.988	76	14	5.997	65¾	24¼										5.991	45½	44½	5.995	42	48	5.994	39	51

LEADS FROM 6.000 TO 6.600

Approximate Lead	Gear on Worm 56, 1st on Stud 32, 2nd on Stud 40, Gear on Screw 100	Angle to Set Spiral Head (Degrees)	Angle to Set Vertical Att. (Degrees)	Gear on Worm 56, 1st on Stud 28, 2nd on Stud 32, Gear on Screw 86	Angle to Set Spiral Head (Degrees)	Angle to Set Vertical Att. (Degrees)	Gear on Worm 72, 1st on Stud 40, 2nd on Stud 44, Gear on Screw 100	Angle to Set Spiral Head (Degrees)	Angle to Set Vertical Att. (Degrees)	Gear on Worm 56, 1st on Stud 32, 2nd on Stud 48, Gear on Screw 100	Angle to Set Spiral Head (Degrees)	Angle to Set Vertical Att. (Degrees)	Gear on Worm 86, 1st on Stud 32, 2nd on Stud 24, Gear on Screw 72	Angle to Set Spiral Head (Degrees)	Angle to Set Vertical Att. (Degrees)	Gear on Worm 64, 1st on Stud 48, 2nd on Stud 40, Gear on Screw 56	Angle to Set Spiral Head (Degrees)	Angle to Set Vertical Att. (Degrees)	Gear on Worm 100, 1st on Stud 44, 2nd on Stud 32, Gear on Screw 72	Angle to Set Spiral Head (Degrees)	Angle to Set Vertical Att. (Degrees)	Gear on Worm 86, 1st on Stud 40, 2nd on Stud 32, Gear on Screw 64	Angle to Set Spiral Head (Degrees)	Angle to Set Vertical Att. (Degrees)
6.000-20	6.000	59	31	6.002	53¾	36¼	6.000	49¼	40¾	6.017	45¾	44¼							6.008	36½	53½	·6.011	34	56
6.020-40	6.031	59½	30½	6.021	54	36	6.022	49½	40½				6.024	42¼	47¾	6.026	39¼	50¾						
6.040-60	6.047	59¾	30¼	6.040	54¼	35¾	6.044	49¾	40¼	6.043	46	44	6.052	42½	47½	6.058	39½	50½	6.044	36¾	53¼	6.050	34¼	55¾
6.060-80	6.062	60	30	6.077	54¾	35¼	6.067	50	40	6.068	46¼	43¾							6.079	37	53			
6.080-00	6.093	60½	29½	6.096	55	35	6.089	50¼	39¾	6.093	46½	43½	6.081	42¾	47¼	6.090	39¾	50¼				6.089	34½	55½
6.100-20	6.108	60¾	29¼	6.115	55¼	34¾	6.111	50½	39½	6.118	46¾	43¼	6.110	43	47				6.114	37¼	52¾			
6.120-40	6.122	61	29	6.134	55½	34½	6.133	50¾	39¼				6.138	43¼	46¾	6.122	40	50				6.128	34¾	55¼
6.140-60	6.152	61½	28½	6.152	55¾	34¼	6.155	51	39	6.143	47	43				6.154	40¼	49¾	6.149	37½	52½			
6.160-80	6.166	61¾	28¼	6.170	56	34	6.177	51¼	38¾	6.168	47¼	42¾	6.167	43½	46½							6.166	35	55
6.180-00	6.181	62	28	6.188	56¼	33¾	6.198	51½	38½	6.193	47½	42½	6.195	43¾	46¼	6.185	40½	49½	6.184	37¾	52¼			
6.200-20	6.209	62½	27½	6.206	56½	33½				6.218	47¾	42¼				6.217	40¾	49¼	6.219	38	52	6.204	35¼	54¾
6.220-40	6.237	63	27	6.224	56¾	33¼	6.220	51¾	38¼				6.223	44	46									
6.240-60	6.251	63¼	26¾	6.242	57	33	6.241	52	38	6.242	48	42	6.251	44¼	45¾	6.248	41	49	6.254	38¼	51¾	6.243	35½	54½
6.260-80	6.265	63½	26½	6.276	57½	32½	6.262	52¼	37¾	6.267	48¼	41¾	6.279	44½	45½									
6.280-00	6.292	64	26	6.294	57¾	32¼	6.283	52½	37½	6.291	48½	41½				6.280	41¼	48¾	6.288	38½	51½	6.281	35¾	54¼
6.300-20	6.318	64½	25½	6.312	58	32	6.304	52¾	37¼	6.315	48¾	41¼	6.307	44¾	45¼	6.311	41½	48½				6.319	36	54
6.320-40	6.331	64¾	25¼	6.329	58¼	31¾	6.325	53	37				6.335	45	45				6.322	38¾	51¼			
6.340-60	6.344	65	25	6.346	58½	31½	6.346	53¼	36¾	6.340	49	41				6.342	41¾	48¼	6.357	39	51	6.357	36¼	53¾
6.360-80	6.370	65½	24½	6.379	59	31	6.367	53½	36½	6.364	49¼	40¾	6.362	45¼	44¾	6.373	42	48						
6.380-00	6.395	66	24	6.396	59¼	30¾	6.387	53¾	36¼	6.387	49½	40½	6.390	45½	44½				6.391	39¼	50¾	6.395	36½	53½
6.400-20	6.419	66½	23½	6.412	59½	30½	6.407	54	36	6.411	49¾	40¼	6.417	45¾	44¼	6.404	42¼	47¾						
6.420-40	6.432	66¾	23¼	6.429	59¾	30¼	6.428	54¼	35¾	6.435	50	40				6.434	42½	47½	6.425	39½	50½	6.432	36¾	53¼
6.440-60	6.444	67	23	6.445	60	30	6.448	54½	35½	6.458	50¼	39¾	6.444	46	44				6.459	39¾	50¼			
6.460-80	6.468	67½	22½	6.477	60½	29½	6.468	54¾	35¼				6.471	46¼	43¾	6.465	42¾	47¼				6.470	37	53
6.480-00	6.490	68	22	6.493	60¾	29¼	6.488	55	35	6.482	50½	39½	6.498	46½	43½	6.495	43	47	6.493	40	50			
6.500-20	6.513	68½	21½	6.509	61	29	6.507	55¼	34¾	6.505	50¾	39¼										6.507	37¼	52¾
6.520-40	6.535	69	21	6.525	61¼	28¾	6.527	55½	34½	6.528	51	39	6.525	46¾	43¼	6.526	43¼	46¾	6.526	40¼	49¾			
6.540-60	6.557	69½	20½	6.540	61½	28½	6.547	55¾	34¼	6.551	51¼	38¾	6.551	47	43	6.556	43½	46½				6.544	37½	52½
6.560-80	6.578	70	20	6.571	62	28	6.566	56	34	6.574	51½	38½	6.578	47¼	42¾				6.560	40½	49½			
6.580-00	6.599	70½	19½	6.586	62¼	27¾	6.585	56¼	33¾	6.597	51¾	38¼				6.586	43¾	46¼	6.594	40¾	49¼	6.582	37¾	52¼

Approximate Lead	Gear on Worm 56, 1st on Stud 28, 2nd on Stud 32, Gear on Screw 86	Angle to Set Spiral Head (Degrees)	Angle to Set Vertical Att. (Degrees)	Gear on Worm 72, 1st on Stud 40, 2nd on Stud 44, Gear on Screw 100	Angle to Set Spiral Head (Degrees)	Angle to Set Vertical Att. (Degrees)	Gear on Worm 56, 1st on Stud 32, 2nd on Stud 48, Gear on Screw 100	Angle to Set Spiral Head (Degrees)	Angle to Set Vertical Att. (Degrees)	Gear on Worm 86, 1st on Stud 32, 2nd on Stud 24, Gear on Screw 72	Angle to Set Spiral Head (Degrees)	Angle to Set Vertical Att. (Degrees)	Gear on Worm 64, 1st on Stud 48, 2nd on Stud 40, Gear on Screw 56	Angle to Set Spiral Head (Degrees)	Angle to Set Vertical Att. (Degrees)	Gear on Worm 100, 1st on Stud 44, 2nd on Stud 32, Gear on Screw 72	Angle to Set Spiral Head (Degrees)	Angle to Set Vertical Att. (Degrees)	Gear on Worm 86, 1st on Stud 40, 2nd on Stud 32, Gear on Screw 64	Angle to Set Spiral Head (Degrees)	Angle to Set Vertical Att. (Degrees)	Gear on Worm 86, 1st on Stud 48, 2nd on Stud 64, Gear on Screw 100	Angle to Set Spiral Head (Degrees)	Angle to Set Vertical Att. (Degrees)
6.600-20	6.601	62½	27½	6.604	56½	33½	6.619	52	38	6.604	47½	42½	6.616	44	46				6.619	38	52	6.618	35¼	54¾
6.620-40	6.631	63	27	6.623	56¾	33¼				6.631	47¾	42¼				6.627	41	49						
6.640-60	6.645	63¼	26¾	6.642	57	33	6.642	52¼	37¾	6.657	48	42	6.646	44¼	45¾				6.655	38¼	51¾	6.659	35½	54½
6.660-80	6.660	63½	26½	6.661	57¼	32¾	6.664	52½	37½				6.675	44½	45½	6.660	41¼	48¾						
6.680-00	6.689	64	26	6.680	57½	32½	6.686	52¾	37¼	6.684	48¼	41¾				6.693	41½	48½	6.692	38½	51½			
6.700-20	6.717	64½	25½	6.717	58	32	6.709	53	37	6.710	48½	41½	6.705	44¾	45¼							6.700	35¾	54¼
6.720-40	6.731	64¾	25¼	6.735	58¼	31¾	6.731	53¼	36¾	6.735	48¾	41¼	6.734	45	45	6.726	41¾	48¼	6.728	38¾	51¼			
6.740-60	6.745	65	25	6.753	58½	31½	6.752	53½	36½							6.759	42	48				6.740	36	54
6.760-80	6.772	65½	24½	6.771	58¾	31¼	6.774	53¾	36¼	6.761	49	41	6.764	45¼	44¾				6.766	39	51			
6.780-00	6.799	66	24	6.789	59	31	6.796	54	36	6.786	49¼	40¾	6.793	45½	44½	6.792	42¼	47¾				6.781	36¼	53¾
6.800-20	6.812	66¼	23¾	6.807	59¼	30¾	6.817	54¼	35¾	6.812	49½	40½							6.802	39¼	50¾			
6.820-40	6.825	66½	23½	6.824	59½	30½	6.839	54½	35½	6.837	49¾	40¼	6.822	45¾	44¼	6.824	42½	47½	6.838	39½	50½	6.821	36½	53½
6.840-60	6.850	67	23	6.859	60	30							6.851	46	44	6.856	42¾	47¼						
6.860-80	6.875	67½	22½	6.876	60¼	29¾	6.860	54¾	35¼	6.862	50	40							6.874	39¾	50¼	6.861	36¾	53¼
6.880-00	6.888	67¾	22¼	6.893	60½	29½	6.881	55	35	6.887	50¼	39¾	6.880	46¼	43¾	6.889	43	47						
6.900-20	6.900	68	22	6.910	60¾	29¼	6.902	55¼	34¾	6.912	50½	39½	6.908	46½	43½				6.910	40	50	6.901	37	53
6.920-40	6.924	68½	21½	6.927	61	29	6.923	55½	34½	6.937	50¾	39¼	6.937	46¾	43¼	6.921	43¼	46¾						
6.940-60	6.948	69	21	6.944	61¼	28¾	6.943	55¾	34¼							6.953	43½	46½	6.946	40¼	49¾	6.941	37¼	52¾
6.960-80	6.971	69½	20½	6.960	61½	28½	6.964	56	34	6.962	51	39	6.965	47	43									
6.980-00	6.993	70	20	6.993	62	28	6.984	56¼	33¾	6.986	51¼	38¾	6.993	47¼	42¾	6.985	43¾	46¼	6.981	40½	49½	6.981	37½	52½
7.000-20	7.015	70½	19½	7.009	62¼	27¾	7.005	56½	33½	7.011	51½	38½				7.017	44	46	7.017	40¾	49¼			
7.020-40	7.036	71	19	7.025	62½	27½	7.025	56¾	33¼	7.035	51¾	38¼	7.022	47½	42½							7.020	37¾	52¼
7.040-60	7.057	71½	18½	7.057	63	27	7.045	57	33	7.059	52	38	7.050	47¾	42¼	7.048	44¼	45¾	7.052	41	49			
7.060-80	7.078	72	18	7.072	63¼	26¾	7.065	57¼	32¾				7.078	48	42							7.060	38	52
7.080-00	7.097	72½	17½	7.088	63½	26½	7.085	57½	32½	7.083	52¼	37¾				7.080	44½	45½	7.088	41¼	48¾	7.099	38¼	51¾
7.100-20	7.117	73	17	7.118	64	26	7.104	57¾	32¼	7.107	52½	37½	7.105	48¼	41¾	7.111	44¾	45¼						
7.120-40	7.135	73½	16½	7.134	64¼	25¾	7.124	58	32	7.131	52¾	37¼	7.133	48½	41½				7.124	41½	48½	7.138	38½	51½
7.140-60	7.154	74	16	7.149	64½	25½	7.143	58¼	31¾	7.155	53	37				7.143	45	45	7.158	41¾	48¼			
7.160-80	7.171	74½	15½	7.178	65	25	7.162	58½	31½	7.178	53¼	36¾	7.160	48¾	41¼	7.174	45¼	44¾				7.177	38¾	51¼
7.180-00	7.188	75	15	7.192	65¼	24¾	7.181	58¾	31¼				7.188	49	41				7.193	42	48			

LEADS FROM 7.200 TO 7.800

Approximate Lead	Gear on Worm 72, 1st on Stud 40, 2nd on Stud 44, Gear on Screw 100	Angle to Set Spiral Head (Degrees)	Angle to Set Vertical Att. (Degrees)	Gear on Worm 56, 1st on Stud 32, 2nd on Stud 48, Gear on Screw 100	Angle to Set Spiral Head (Degrees)	Angle to Set Vertical Att. (Degrees)	Gear on Worm 86, 1st on Stud 32, 2nd on Stud 24, Gear on Screw 72	Angle to Set Spiral Head (Degrees)	Angle to Set Vertical Att. (Degrees)	Gear on Worm 64, 1st on Stud 48, 2nd on Stud 40, Gear on Screw 56	Angle to Set Spiral Head (Degrees)	Angle to Set Vertical Att. (Degrees)	Gear on Worm 100, 1st on Stud 44, 2nd on Stud 32, Gear on Screw 72	Angle to Set Spiral Head (Degrees)	Angle to Set Vertical Att. (Degrees)	Gear on Worm 86, 1st on Stud 40, 2nd on Stud 32, Gear on Screw 64	Angle to Set Spiral Head (Degrees)	Angle to Set Vertical Att. (Degrees)	Gear on Worm 86, 1st on Stud 48, 2nd on Stud 64, Gear on Screw 100	Angle to Set Spiral Head (Degrees)	Angle to Set Vertical Att. (Degrees)	Gear on Worm 72, 1st on Stud 44, 2nd on Stud 64, Gear on Screw 86	Angle to Set Spiral Head (Degrees)	Angle to Set Vertical Att. (Degrees)
7.200-20	7.207	65½	24½	7.219	59¼	30¾	7.201	53½	36½	7.215	49¼	40¾	7.204	45½	44½				7.216	39	51	7.201	36¼	53¾
7.220-40	7.235	66	24	7.238	59½	30½	7.224	53¾	36¼				7.235	45¾	44¼	7.228	42¼	47¾						
7.240-60	7.249	66¼	23¾	7.256	59¾	30¼	7.248	54	36	7.242	49½	40½							7.255	39¼	50¾	7.244	36½	53½
7.260-80	7.263	66½	23½	7.275	60	30	7.270	54¼	35¾	7.269	49¾	40¼	7.266	46	44	7.263	42½	47½						
7.280-00	7.290	67	23	7.293	60¼	29¾	7.293	54½	35½	7.296	50	40	7.297	46¼	43¾	7.297	42¾	47¼	7.294	39½	50½	7.286	36¾	53¼
7.300-20	7.304	67¼	22¾	7.311	60½	29½	7.316	54¾	35¼															
7.320-40	7.330	67¾	22¼	7.329	60¾	29¼	7.338	55	35	7.322	50¼	39¾	7.327	46½	43½	7.332	43	47	7.332	39¾	50¼	7.329	37	53
7.340-60	7.343	68	22	7.347	61	29				7.349	50½	39½	7.357	46¾	43¼									
7.360-80	7.369	68½	21½	7.365	61¼	28¾	7.361	55¼	34¾	7.375	50¾	39¼				7.366	43¼	46¾	7.370	40	50	7.371	37¼	52¾
7.380-00	7.394	69	21	7.382	61½	28½	7.383	55½	34½				7.387	47	43									
7.400-20	7.418	69½	20½	7.417	62	28	7.405	55¾	34¼	7.401	51	39	7.418	47¼	42¾	7.400	43½	46½	7.408	40¼	49¾	7.413	37½	52½
7.420-40	7.430	69¾	20¼	7.434	62¼	27¾	7.427	56	34	7.427	51¼	38¾				7.434	43¾	46¼						
7.440-60	7.442	70	20	7.451	62½	27½	7.449	56¼	33¾	7.453	51½	38½	7.447	47½	42½				7.446	40½	49½	7.455	37¾	52¼
7.460-80	7.466	70½	19½	7.468	62¾	27¼	7.471	56½	33½	7.479	51¾	38¼	7.477	47¾	42¼	7.467	44	46						
7.480-00	7.489	71	19	7.484	63	27	7.492	56¾	33¼										7.484	40¾	49¼	7.497	38	52
7.500-20	7.511	71½	18½	7.517	63½	26½	7.513	57	33	7.505	52	38	7.506	48	42	7.501	44¼	45¾						
7.520-40	7.532	72	18	7.534	63¾	26¼	7.534	57¼	32¾	7.530	52¼	37¾	7.536	48¼	41¾	7.535	44½	45½	7.522	41	49	7.539	38¼	51¾
7.540-60	7.543	72¼	17¾	7.550	64	26	7.555	57½	32½	7.556	52½	37½												
7.560-80	7.574	73	17	7.566	64¼	25¾	7.576	57¾	32¼				7.565	48½	41½	7.568	44¾	45¼	7.560	41¼	48¾			
7.580-00	7.594	73½	16½	7.582	64½	25½	7.597	58	32	7.581	52¾	37¼	7.594	48¾	41¼				7.598	41½	48½	7.580	38½	51½
7.600-20	7.613	74	16	7.613	65	25	7.618	58¼	31¾	7.606	53	37				7.602	45	45						
7.620-40	7.632	74½	15½	7.628	65¼	24¾	7.638	58½	31½	7.631	53¼	36¾	7.624	49	41	7.634	45¼	44¾	7.636	41¾	48¼	7.622	38¾	51¼
7.640-60	7.650	75	15	7.644	65½	24½	7.659	58¾	31¼	7.656	53½	36½	7.652	49¼	40¾									
7.660-80	7.668	75½	14½	7.674	66	24	7.679	59	31							7.668	45½	44½	7.672	42	48	7.664	39	51
7.680-00	7.685	76	14	7.689	66¼	23¾	7.699	59¼	30¾	7.680	53¾	36¼	7.681	49½	40½									
7.700-20	7.717	77	13	7.703	66½	23½	7.719	59½	30½	7.705	54	36	7.710	49¾	40¼	7.700	45¾	44¼	7.710	42¼	47¾	7.704	39¼	50¾
7.720-40	7.725	77¼	12¾	7.732	67	23	7.739	59¾	30¼	7.729	54¼	35¾	7.738	50	40	7.733	46	44						
7.740-60	7.747	78	12	7.746	67¼	22¾	7.758	60	30	7.754	54½	35½							7.746	42½	47½	7.746	39½	50½
7.760-80	7.775	79	11	7.760	67½	22½	7.778	60¼	29¾	7.778	54¾	35¼	7.766	50¼	39¾	7.765	46¼	43¾						
7.780-00	7.787	79½	10½	7.788	68	22	7.797	60½	29½				7.794	50½	39½	7.798	46½	43½	7.783	42¾	47¼	7.787	39¾	50¼

Approximate Lead	Gear on Worm 86, 1st on Stud 32, 2nd on Stud 24, Gear on Screw 72	Angle to Set Spiral Head (Degrees)	Angle to Set Vertical Att. (Degrees)	Gear on Worm 64, 1st on Stud 48, 2nd on Stud 40, Gear on Screw 56	Angle to Set Spiral Head (Degrees)	Angle to Set Vertical Att. (Degrees)	Gear on Worm 100, 1st on Stud 44, 2nd on Stud 32, Gear on Screw 72	Angle to Set Spiral Head (Degrees)	Angle to Set Vertical Att. (Degrees)	Gear on Worm 86, 1st on Stud 40, 2nd on Stud 32, Gear on Screw 64	Angle to Set Spiral Head (Degrees)	Angle to Set Vertical Att. (Degrees)	Gear on Worm 86, 1st on Stud 48, 2nd on Stud 64, Gear on Screw 100	Angle to Set Spiral Head (Degrees)	Angle to Set Vertical Att. (Degrees)	Gear on Worm 72, 1st on Stud 44, 2nd on Stud 64, Gear on Screw 86	Angle to Set Spiral Head (Degrees)	Angle to Set Vertical Att. (Degrees)	Gear on Worm 56, 1st on Stud 24, 2nd on Stud 40, Gear on Screw 72	Angle to Set Spiral Head (Degrees)	Angle to Set Vertical Att. (Degrees)	Gear on Worm 86, 1st on Stud 40, 2nd on Stud 64, Gear on Screw 100	Angle to Set Spiral Head (Degrees)	Angle to Set Vertical Att. (Degrees)
7.800-20	7.816	60¾	29¼	7.801	55	35													7.802	37	53			
7.820-40	7.835	61	29	7.825	55¼	34¾	7.822	50¾	39¼	7.830	46¾	43¼	7.820	43	47	7.827	40	50						
7.840-60	7.854	61¼	28¾	7.849	55½	34½	7.850	51	39				7.856	43¼	46¾				7.846	37¼	52¾	7.843	34¾	55¼
7.860-80	7.873	61½	28½	7.872	55¾	34¼	7.877	51¼	38¾	7.862	47	43				7.868	40¼	49¾						
7.880-00	7.891	61¾	28¼	7.896	56	34				7.894	47¼	42¾	7.893	43½	46½				7.891	37½	52½	7.892	35	55
7.900-20	7.910	62	28	7.919	56¼	33¾	7.905	51½	38½							7.908	40½	49½						
7.920-40	7.928	62¼	27¾				7.932	51¾	38¼	7.926	47½	42½	7.928	43¾	46¼				7.936	37¾	52¼			
7.940-60	7.946	62½	27½	7.942	56½	33½				7.958	47¾	42¼				7.948	40¾	49¼				7.942	35¼	54¾
7.960-80	7.964	62¾	27¼	7.965	56¾	33¼	7.960	52	38				7.965	44	46									
7.980-00	7.982	63	27	7.987	57	33	7.987	52¼	37¾	7.989	48	42				7.989	41	49	7.981	38	52	7.990	35½	54½
8.000-20	8.017	63½	26½	8.010	57¼	32¾	8.014	52½	37½				8.001	44¼	45¾									
8.020-40	8.034	63¾	26¼	8.032	57½	32½				8.020	48¼	41¾	8.037	44½	45½	8.029	41¼	48¾	8.026	38¼	51¾			
8.040-60	8.052	64	26	8.055	57¾	32¼	8.040	52¾	37¼	8.052	48½	41½										8.040	35¾	54¼
8.060-80	8.069	64¼	25¾	8.077	58	32	8.067	53	37				8.072	44¾	45¼	8.069	41½	48½	8.070	38½	51½			
8.080-00	8.086	64½	25½	8.099	58¼	31¾	8.094	53¼	36¾	8.082	48¾	41¼										8.088	36	54
8.100-20	8.119	65	25							8.113	49	41	8.108	45	45	8.109	41¾	48¼	8.114	38¾	51¼			
8.120-40	8.136	65¼	24¾	8.120	58½	31½	8.120	53½	36½													8.137	36¼	53¾
8.140-60	8.152	65½	24½	8.142	58¾	31¼	8.146	53¾	36¼	8.144	49¼	40¾	8.143	45¼	44¾	8.148	42	48	8.158	39	51			
8.160-80	8.168	65¾	24¼	8.163	59	31	8.172	54	36	8.174	49½	40½	8.178	45½	44½									
8.180-00	8.184	66	24	8.185	59¼	30¾	8.198	54¼	35¾							8.188	42¼	47¾				8.185	36½	53½
8.200-20	8.216	66½	23½	8.206	59½	30½				8.205	49¾	40¼	8.213	45¾	44¼				8.202	39¼	50¾			
8.220-40	8.230	66¾	23¼	8.227	59¾	30¼	8.224	54½	35½	8.235	50	40				8.226	42½	47½				8.233	36¾	53¼
8.240-60	8.246	67	23	8.248	60	30	8.249	54¾	35¼				8.248	46	44				8.246	39½	50½			
8.260-80	8.277	67½	22½	8.269	60¼	29¾	8.274	55	35	8.265	50¼	39¾				8.266	42¾	47¼						
8.280-00	8.292	67¾	22¼	8.289	60½	29½				8.295	50½	39½	8.282	46¼	43¾				8.289	39¾	50¼	8.281	37	53
8.300-20	8.306	68	22	8.310	60¾	29¼	8.300	55¼	34¾				8.317	46½	43½	8.305	43	47						
8.320-40	8.335	68½	21½	8.330	61	29	8.324	55½	34½	8.324	50¾	39¼							8.332	40	50	8.329	37¼	52¾
8.340-60	8.350	68¾	21¼	8.350	61¼	28¾	8.349	55¾	34¼	8.354	51	39	8.351	46¾	43¼	8.343	43¼	46¾						
8.360-80	8.364	69	21	8.370	61½	28½	8.374	56	34										8.375	40¼	49¾	8.376	37½	52½
8.380-00	8.392	69½	20½	8.389	61¾	28¼	8.398	56¼	33¾	8.384	51¼	38¾	8.386	47	43	8.382	43½	46½						

LEADS FROM 8.400 TO 9.000

Approximate Lead	Gear on Worm 64, 1st on Stud 48, 2nd on Stud 40, Gear on Screw 56	Angle to Set Spiral Head (Degrees)	Angle to Set Vertical Att. (Degrees)	Gear on Worm 100, 1st on Stud 44, 2nd on Stud 32, Gear on Screw 72	Angle to Set Spiral Head (Degrees)	Angle to Set Vertical Att. (Degrees)	Gear on Worm 86, 1st on Stud 40, 2nd on Stud 32, Gear on Screw 64	Angle to Set Spiral Head (Degrees)	Angle to Set Vertical Att. (Degrees)	Gear on Worm 86, 1st on Stud 48, 2nd on Stud 64, Gear on Screw 100	Angle to Set Spiral Head (Degrees)	Angle to Set Vertical Att. (Degrees)	Gear on Worm 72, 1st on Stud 44, 2nd on Stud 64, Gear on Screw 86	Angle to Set Spiral Head (Degrees)	Angle to Set Vertical Att. (Degrees)	Gear on Worm 56, 1st on Stud 24, 2nd on Stud 40, Gear on Screw 72	Angle to Set Spiral Head (Degrees)	Angle to Set Vertical Att. (Degrees)	Gear on Worm 86, 1st on Stud 40, 2nd on Stud 64, Gear on Screw 100	Angle to Set Spiral Head (Degrees)	Angle to Set Vertical Att. (Degrees)	Gear on Worm 86, 1st on Stud 44, 2nd on Stud 48, Gear on Screw 64	Angle to Set Spiral Head (Degrees)	Angle to Set Vertical Att. (Degrees)
8.400-20	8.409	62	28				8.413	51½	38½							8.419	40½	49½				8.408	35	55
8.420-40	8.428	62¼	27¾	8.424	56½	33½				8.420	47¼	42¾	8.420	43¾	46¼				8.424	37¾	52¼			
8.440-60	8.448	62½	27½	8.448	56¾	33¼	8.443	51¾	38¼	8.454	47½	42½	8.459	44	46									
8.460-80	8.467	62¾	27¼	8.472	57	33	8.471	52	38							8.462	40¾	49¼	8.471	38	52	8.460	35¼	54¾
8.480-00	8.486	63	27	8.496	57¼	32¾				8.488	47¾	42¼	8.497	44¼	45¾									
8.500-20	8.504	63¼	26¾	8.519	57½	32½	8.500	52¼	37¾							8.505	41	49	8.519	38¼	51¾	8.513	35½	54½
8.520-40	8.523	63½	26½				8.529	52½	37½	8.521	48	42	8.536	44½	45½									
8.540-60	8.542	63¾	26¼	8.543	57¾	32¼	8.557	52¾	37¼	8.555	48¼	41¾				8.547	41¼	48¾						
8.560-80	8.560	64	26	8.566	58	32							8.573	44¾	45¼				8.566	38½	51½	8.565	35¾	54¼
8.580-00	8.596	64½	25½	8.590	58¼	31¾	8.585	53	37	8.588	48½	41½				8.590	41½	48½						
8.600-20	8.614	64¾	25¼	8.612	58½	31½	8.614	53¼	36¾				8.611	45	45				8.613	38¾	51¼	8.618	36	54
8.620-40	8.632	65	25	8.636	58¾	31¼				8.621	48¾	41¼				8.632	41¾	48¼						
8.640-60	8.649	65¼	24¾	8.658	59	31	8.642	53½	36½	8.654	49	41	8.648	45¼	44¾									
8.660-80	8.666	65½	24½				8.670	53¾	36¼							8.674	42	48	8.660	39	51	8.669	36¼	53¾
8.680-00	8.684	65¾	24¼	8.681	59¼	30¾	8.697	54	36	8.686	49¼	40¾	8.686	45½	44½									
8.700-20	8.701	66	24	8.703	59½	30½				8.719	49½	40½				8.716	42¼	47¾	8.706	39¼	50¾			
8.720-40	8.734	66½	23½	8.726	59¾	30¼	8.724	54¼	35¾				8.723	45¾	44¼							8.720	36½	53½
8.740-60	8.750	66¾	23¼	8.748	60	30	8.752	54½	35½	8.751	49¾	40¼				8.758	42½	47½	8.752	39½	50½			
8.760-80	8.766	67	23	8.770	60¼	29¾	8.779	54¾	35¼				8.760	46	44							8.771	36¾	53¼
8.780-00	8.799	67½	22½	8.791	60½	29½				8.784	50	40	8.796	46¼	43¾	8.799	42¾	47¼	8.799	39¾	50¼			
8.800-20	8.815	67¾	22¼	8.813	60¾	29¼	8.806	55	35	8.815	50¼	39¾												
8.820-40	8.830	68	22	8.834	61	29	8.832	55¼	34¾				8.833	46½	43½							8.822	37	53
8.840-60	8.846	68¼	21¾	8.856	61¼	28¾				8.847	50½	39½				8.841	43	47	8.844	40	50			
8.860-80	8.861	68½	21½	8.877	61½	28½	8.860	55½	34½	8.879	50¾	39¼	8.869	46¾	43¼							8.873	37¼	52¾
8.880-00	8.891	69	21	8.898	61¾	28¼	8.886	55¾	34¼							8.882	43¼	46¾	8.890	40¼	49¾			
8.900-20	8.906	69¼	20¾	8.919	62	28	8.912	56	34	8.910	51	39	8.906	47	43									
8.920-40	8.921	69½	20½				8.938	56¼	33¾							8.924	43½	46½	8.936	40½	49½	8.924	37½	52½
8.940-60	8.949	70	20	8.940	62¼	27¾				8.942	51¼	38¾	8.942	47¼	42¾									
8.960-80	8.978	70½	19½	8.960	62½	27½	8.965	56½	33½	8.974	51½	38½	8.978	47½	42½	8.964	43¾	46¼				8.975	37¾	52¼
8.980-00	8.991	70¾	19¼	8.980	62¾	27¼	8.990	56¾	33¼										8.982	40¾	49¼			

Approximate Lead	Gear on Worm 100, 1st on Stud 44, 2nd on Stud 32, Gear on Screw 72	Angle to Set Spiral Head (Degrees)	Angle to Set Vertical Att. (Degrees)	Gear on Worm 86, 1st on Stud 40, 2nd on Stud 32, Gear on Screw 64	Angle to Set Spiral Head (Degrees)	Angle to Set Vertical Att. (Degrees)	Gear on Worm 86, 1st on Stud 48, 2nd on Stud 64, Gear on Screw 100	Angle to Set Spiral Head (Degrees)	Angle to Set Vertical Att. (Degrees)	Gear on Worm 72, 1st on Stud 44, 2nd on Stud 64, Gear on Screw 86	Angle to Set Spiral Head (Degrees)	Angle to Set Vertical Att. (Degrees)	Gear on Worm 56, 1st on Stud 24, 2nd on Stud 40, Gear on Screw 72	Angle to Set Spiral Head (Degrees)	Angle to Set Vertical Att. (Degrees)	Gear on Worm 86, 1st on Stud 40, 2nd on Stud 64, Gear on Screw 100	Angle to Set Spiral Head (Degrees)	Angle to Set Vertical Att. (Degrees)	Gear on Worm 86, 1st on Stud 44, 2nd on Stud 48, Gear on Screw 64	Angle to Set Spiral Head (Degrees)	Angle to Set Vertical Att. (Degrees)	Gear on Worm 64, 1st on Stud 32, 2nd on Stud 56, Gear on Screw 72	Angle to Set Spiral Head (Degrees)	Angle to Set Vertical Att. (Degrees)
9.000-20	9.000	63	27	9.016	57	33	9.004	51¾	38¼	9.014	47¾	42¼	9.005	44	46									
9.020-40	9.020	63¼	26¾				9.036	52	38							9.028	41	49	9.025	38	52	9.034	35½	54½
9.040-60	9.040	63½	26½	9.042	57¼	32¾				9.050	48	42	9.046	44¼	45¾									
9.060-80	9.079	64	26	9.067	57½	32½	9.067	52¼	37¾							9.073	41¼	48¾	9.076	38¼	51¾			
9.080-00	9.098	64¼	25¾	9.092	57¾	32¼	9.097	52½	37½	9.086	48¼	41¾	9.086	44½	45½							9.089	35¾	54¼
9.100-20	9.117	64½	25½	9.117	58	32										9.118	41½	48½						
9.120-40	9.136	64¾	25¼				9.127	52¾	37¼	9.120	48½	41½	9.126	44¾	45¼				9.126	38½	51½			
9.140-60	9.154	65	25	9.141	58¼	31¾	9.157	53	37	9.156	48¾	41¼										9.144	36	54
9.160-80	9.173	65¼	24¾	9.166	58½	31½							9.167	45	45	9.163	41¾	48¼	9.175	38¾	51¼			
9.180-00	9.191	65½	24½	9.190	58¾	31¼	9.187	53¼	36¾	9.190	49	41										9.199	36¼	53¾
9.200-20	9.210	65¾	24¼	9.214	59	31	9.217	53½	36½				9.206	45¼	44¾	9.207	42	48						
9.220-40	9.228	66	24	9.238	59¼	30¾				9.225	49¼	40¾							9.225	39	51			
9.240-60	9.245	66¼	23¾				9.247	53¾	36¼				9.246	45½	44½	9.252	42¼	47¾				9.254	36½	53½
9.260-80	9.263	66½	23½	9.262	59½	30½	9.276	54	36	9.260	49½	40½							9.275	39¼	50¾			
9.280-00	9.298	67	23	9.286	59¾	30¼				9.294	49¾	40¼	9.286	45¾	44¼	9.296	42½	47½						
9.300-20	9.315	67¼	22¾	9.310	60	30	9.305	54¼	35¾													9.308	36¾	53¼
9.320-40	9.332	67½	22½	9.333	60¼	29¾	9.334	54½	35½	9.328	50	40	9.325	46	44				9.324	39½	50½			
9.340-60	9.349	67¾	22¼	9.356	60½	29½										9.340	42¾	47¼						
9.360-80	9.365	68	22	9.379	60¾	29¼	9.364	54¾	35¼	9.362	50¼	39¾	9.364	46¼	43¾				9.374	39¾	50¼	9.362	37	53
9.380-00	9.398	68½	21½				9.392	55	35	9.396	50½	39½				9.384	43	47						
9.400-20	9.414	68¾	21¼	9.402	61	29							9.403	46½	43½							9.416	37¼	52¾
9.420-40	9.430	69	21	9.425	61¼	28¾	9.421	55¼	34¾	9.430	50¾	39¼				9.428	43¼	46¾	9.422	40	50			
9.440-60	9.446	69¼	20¾	9.448	61½	28½	9.450	55½	34½				9.442	46¾	43¼									
9.460-80	9.462	69½	20½	9.470	61¾	28¼	9.478	55¾	34¼	9.464	51	39				9.472	43½	46½	9.471	40¼	49¾	9.470	37½	52½
9.480-00	9.492	70	20	9.492	62	28				9.497	51¼	38¾	9.481	47	43									
9.500-20	9.507	70¼	19¾	9.514	62¼	27¾	9.506	56	34							9.515	43¾	46¼						
9.520-40	9.522	70½	19½	9.536	62½	27½	9.534	56¼	33¾	9.530	51½	38½	9.520	47¼	42¾				9.520	40½	49½	9.524	37¾	52¼
9.540-60	9.550	71	19	9.548	62¾	27¼							9.548	47½	42½	9.559	44	46						
9.560-80	9.579	71½	18½	9.579	63	27	9.562	56½	33½	9.563	51¾	38¼							9.569	40¾	49¼	9.578	38	52
9.580-00	9.593	71¾	18¼							9.596	52	38	9.596	47¾	42¼									

LEADS FROM 9.600 TO 10.200

Approximate Lead	Gear on Worm 86, 1st on Stud 40, 2nd on Stud 32, Gear on Screw 64	Angle to Set Spiral Head (Degrees)	Angle to Set Vertical Att. (Degrees)	Gear on Worm 86, 1st on Stud 48, 2nd on Stud 64, Gear on Screw 100	Angle to Set Spiral Head (Degrees)	Angle to Set Vertical Att. (Degrees)	Gear on Worm 72, 1st on Stud 44, 2nd on Stud 64, Gear on Screw 86	Angle to Set Spiral Head (Degrees)	Angle to Set Vertical Att. (Degrees)	Gear on Worm 56, 1st on Stud 24, 2nd on Stud 40, Gear on Screw 72	Angle to Set Spiral Head (Degrees)	Angle to Set Vertical Att. (Degrees)	Gear on Worm 86, 1st on Stud 40, 2nd on Stud 64, Gear on Screw 100	Angle to Set Spiral Head (Degrees)	Angle to Set Vertical Att. (Degrees)	Gear on Worm 86, 1st on Stud 44, 2nd on Stud 48, Gear on Screw 64	Angle to Set Spiral Head (Degrees)	Angle to Set Vertical Att. (Degrees)	Gear on Worm 64, 1st on Stud 32, 2nd on Stud 56, Gear on Screw 72	Angle to Set Spiral Head (Degrees)	Angle to Set Vertical Att. (Degrees)	Gear on Worm 72, 1st on Stud 40, 2nd on Stud 44, Gear on Screw 48	Angle to Set Spiral Head (Degrees)	Angle to Set Vertical Att. (Degrees)
9.600-20	9.600	63¼	26¾	9.617	57	33							9.602	44¼	45¾	9.618	41	49						
9.620-40	9.621	63½	26½				9.629	52¼	37¾	9.634	48	42							9.631	38¼	51¾			
9.640-60	9.642	63¾	26¼	9.644	57¼	32¾							9.645	44½	45½							9.641	35¾	54¼
9.660-80	9.662	64	26	9.670	57½	32½	9.661	52½	37½	9.672	48¼	41¾				9.666	41¼	48¾						
9.680-00	9.683	64¼	25¾	9.697	57¾	32¼	9.693	52¾	37¼				9.688	44¾	45¼				9.684	38½	51½	9.699	36	54
9.700-20	9.703	64½	25½							9.710	48½	41½				9.714	41½	48½						
9.720-40	9.723	64¾	25¼	9.724	58	32	9.725	53	37				9.730	45	45				9.736	38¾	51¼			
9.740-60	9.743	65	25	9.750	58¼	31¾	9.757	53¼	36¾	9.747	48¾	41¼										9.757	36¼	53¾
9.760-80	9.763	65¼	24¾	9.776	58½	31½							9.772	45¼	44¾	9.762	41¾	48¼						
9.780-00	9.782	65½	24½				9.788	53½	36½	9.784	49	41							9.790	39	51			
9.800-20	9.801	65¾	24¼	9.802	58¾	31¼							9.814	45½	44½	9.809	42	48				9.815	36½	53½
9.820-40	9.839	66¼	23¾	9.828	59	31	9.820	53¾	36¼	9.820	49¼	40¾												
9.840-60	9.858	66½	23½	9.854	59¼	30¾	9.852	54	36	9.857	49½	40½	9.856	45¾	44¼	9.846	42¼	47¾	9.842	39¼	50¾			
9.860-80	9.876	66¾	23¼	9.879	59½	30½																9.873	36¾	53¼
9.880-00	9.895	67	23				9.882	54¼	35¾	9.894	49¾	40¼	9.898	46	44				9.895	39½	50½			
9.900-20	9.914	67¼	22¾	9.904	59¾	30¼	9.914	54½	35½							9.904	42½	47½						
9.920-40	9.932	67½	22½	9.930	60	30				9.930	50	40										9.930	37	53
9.940-60	9.950	67¾	22¼	9.954	60¼	29¾	9.944	54¾	35¼				9.940	46¼	43¾	9.950	42¾	47¼	9.947	39¾	50¼			
9.960-80	9.967	68	22	9.979	60½	29½	9.975	55	35	9.966	50¼	39¾												
9.980-00	9.985	68¼	21¾										9.981	46½	43½	9.997	43	47	9.999	40	50	9.988	37¼	52¾
10.000-20	10.002	68½	21½	10.004	60¾	29¼	10.006	55¼	34¾	10.002	50½	39½												
10.020-40	10.036	69	21	10.028	61	29	10.036	55½	34½	10.038	50¾	39¼	10.022	46¾	43¼									
10.040-60	10.053	69¼	20¾	10.053	61¼	28¾										10.044	43¼	46¾	10.050	40¼	49¾	10.045	37½	52½
10.060-80	10.070	69½	20½	10.077	61½	28½	10.066	55¾	34¼	10.074	51	39	10.063	47	43									
10.080-00	10.086	69¾	20¼				10.096	56	34							10.091	43½	46½						
10.100-20	10.102	70	20	10.100	61¾	28¼				10.110	51¼	38¾	10.104	47¼	42¾				10.102	40½	49½	10.103	37¾	52¼
10.120-40	10.134	70½	19½	10.124	62	28	10.125	56¼	33¾							10.137	43¾	46¼						
10.140-60	10.149	70¾	19¼	10.148	62¼	27¾	10.155	56½	33½	10.146	51½	38½	10.145	47½	42½				10.154	40¾	49¼	10.159	38	52
10.160-80	10.163	71	19	10.170	62½	27½																		
10.180-00	10.194	71½	18½	10.190	62¾	27¼	10.184	56¾	33¼	10.180	51¾	38¼	10.186	47¾	42¼	10.183	44	46						

Approximate Lead	Gear on Worm 86, 1st on Stud 48, 2nd on Stud 64, Gear on Screw 100	Angle to Set Spiral Head (Degrees)	Angle to Set Vertical Att. (Degrees)	Gear on Worm 72, 1st on Stud 44, 2nd on Stud 64, Gear on Screw 86	Angle to Set Spiral Head (Degrees)	Angle to Set Vertical Att. (Degrees)	Gear on Worm 56, 1st on Stud 24, 2nd on Stud 40, Gear on Screw 72	Angle to Set Spiral Head (Degrees)	Angle to Set Vertical Att. (Degrees)	Gear on Worm 86, 1st on Stud 40, 2nd on Stud 64, Gear on Screw 100	Angle to Set Spiral Head (Degrees)	Angle to Set Vertical Att. (Degrees)	Gear on Worm 86, 1st on Stud 44, 2nd on Stud 48, Gear on Screw 64	Angle to Set Spiral Head (Degrees)	Angle to Set Vertical Att. (Degrees)	Gear on Worm 64, 1st on Stud 32, 2nd on Stud 56, Gear on Screw 72	Angle to Set Spiral Head (Degrees)	Angle to Set Vertical Att. (Degrees)	Gear on Worm 72, 1st on Stud 40, 2nd on Stud 44, Gear on Screw 48	Angle to Set Spiral Head (Degrees)	Angle to Set Vertical Att. (Degrees)	Gear on Worm 100, 1st on Stud 44, 2nd on Stud 56, Gear on Screw 72	A'gle to Set Spiral Head (Degrees)	Angle to Set Vertical Att. (Degrees)
10.200-20	10.217	63	27	10.213	57	33	10.216	52	38							10.206	41	49	10.215	38¼	51¾	10.203	35¼	54¾
10.220-40	10.238	63¼	26¾							10.226	48	42	10.230	44¼	45¾									
10.240-60				10.242	57¼	32¾	10.250	52¼	37¾							10.257	41¼	48¾						
10.260-80	10.260	63½	26½	10.270	57½	32½				10.267	48¼	41¾	10.274	44½	45½				10.272	38½	51½	10.265	35½	54½
10.280-00	10.283	63¾	26¼	10.298	57¾	32¼	10.285	52½	37½															
10.300-20	10.305	64	26				10.319	52¾	37¼	10.306	48½	41½				10.308	41½	48½						
10.320-40	10.327	64¼	25¾	10.327	58	32							10.320	44¾	45¼				10.328	38¾	51¼	10.327	35¾	54¼
10.340-60	10.349	64½	25½	10.354	58¼	31¾	10.353	53	37	10.345	48¾	41¼				10.358	41¾	48¼						
10.360-80	10.370	64¾	25¼										10.366	45	45									
10.380-00	10.391	65	25	10.383	58½	31½	10.387	53¼	36¾	10.384	49	41							10.384	39	51	10.390	36	54
10.400-20	10.412	65¼	24¾	10.410	58¾	31¼							10.410	45¼	44¾	10.409	42	48						
10.420-40	10.433	65½	24½	10.438	59	31	10.420	53½	36½	10.424	49¼	40¾												
10.440-60	10.453	65¾	24¼				10.454	53¾	36¼				10.455	45½	44½				10.440	39¼	50¾	10.453	36¼	53¾
10.460-80	10.474	66	24	10.465	59¼	30¾				10.463	49½	40½				10.460	42¼	47¾						
10.480-00	10.494	66¼	23¾	10.492	59½	30½	10.488	54	36										10.496	39½	50½			
10.500-20	10.515	66½	23½	10.519	59¾	30¼				10.501	49¾	40¼	10.500	45¾	44¼	10.509	42½	47½				10.515	36½	53½
10.520-40	10.534	66¾	23¼				10.520	54¼	35¾															
10.540-60	10.554	67	23	10.546	60	30	10.553	54½	35½	10.540	50	40	10.544	46	44	10.559	42¾	47¼	10.551	39¾	50¼			
10.560-80	10.574	67¼	22¾	10.572	60¼	29¾																10.576	36¾	53¼
10.580-00	10.593	67½	22½	10.598	60½	29½	10.586	54¾	35¼	10.580	50¼	39¾	10.589	46¼	43¾									
10.600-20	10.612	67¾	22¼				10.619	55	35	10.618	50½	39½				10.609	43	47	10.606	40	50			
10.620-40	10.630	68	22	10.625	60¾	29¼							10.633	46½	43½							10.638	37	53
10.640-60	10.650	68¼	21¾	10.650	61	29	10.651	55¼	34¾	10.655	50¾	39¼				10.658	43¼	46¾						
10.660-80	10.668	68½	21½	10.676	61¼	28¾							10.677	46¾	43¼				10.661	40¼	49¾			
10.680-00	10.687	68¾	21¼				10.683	55½	34½	10.693	51	39												
10.700-20	10.704	69	21	10.701	61½	28½	10.716	55¾	34¼							10.708	43½	46½	10.716	40½	49½	10.700	37¼	52¾
10.720-40	10.722	69¼	20¾	10.727	61¾	28¼				10.731	51¼	38¾	10.721	47	43									
10.740-60	10.740	69½	20½	10.752	62	28	10.747	56	34							10.757	43¾	46¼						
10.760-80	10.775	70	20	10.777	62¼	27¾	10.779	56¼	33¾	10.769	51½	38½	10.764	47¼	42¾				10.770	40¾	49¼	10.760	37½	52½
10.780-00	10.791	70¼	19¾																					

LEADS FROM 10.800 TO 11.400

Approximate Lead	Gear on Worm 86, 1st on Stud 48, 2nd on Stud 64, Gear on Screw 100	Angle to Set Spiral Head (Degrees)	Angle to Set Vertical Att. (Degrees)	Gear on Worm 72, 1st on Stud 44, 2nd on Stud 64, Gear on Screw 86	Angle to Set Spiral Head (Degrees)	Angle to Set Vertical Att. (Degrees)	Gear on Worm 56, 1st on Stud 24, 2nd on Stud 40, Gear on Screw 72	Angle to Set Spiral Head (Degrees)	Angle to Set Vertical Att. (Degrees)	Gear on Worm 86, 1st on Stud 40, 2nd on Stud 64, Gear on Screw 100	Angle to Set Spiral Head (Degrees)	Angle to Set Vertical Att. (Degrees)
10.800-20	10.809	70½	19½	10.801	62½	27½	10.810	56½	33½	10.807	51¾	38¼
10.820-40	10.825	70¾	19¼	10.826	62¾	27¼						
10.840-60	10.840	71	19	10.850	63	27	10.841	56¾	33¼	10.843	52	38
10.860-80	10.873	71½	18½	10.875	63¼	26¾	10.872	57	33			
10.880-00	10.889	71¾	18¼	10.898	63½	26½				10.880	52¼	37¾
10.900-20	10.904	72	18				10.903	57¼	32¾	10.917	52½	37½
10.920-40	10.934	72½	17½	10.920	63¾	26¼	10.933	57½	32½			
10.940-60	10.950	72¾	17¼	10.944	64	26				10.953	52¾	37¼
10.960-80	10.964	73	17	10.967	64¼	25¾	10.963	57¾	32¼			
10.980-00	10.992	73½	16½	10.990	64½	25½	10.993	58	32	10.989	53	37
11.000-20	11.007	73¾	16¼	11.013	64¾	25¼						
11.020-40	11.021	74	16	11.036	65	25	11.022	58¼	31¾	11.025	53¼	36¾
11.040-60	11.049	74½	15½	11.058	65¼	24¾	11.052	58½	31½			
11.060-80	11.075	75	15							11.060	53½	36½
11.080-00	11.088	75¼	14¾	11.080	65½	24½	11.082	58¾	31¼	11.097	53¾	36¼
11.100-20	11.100	75½	14½	11.102	65¾	24¼	11.111	59	31			
11.120-40	11.125	75	14	11.124	66	24				11.132	54	36
11.140-60	11.148	76½	13½	11.145	66¼	23¾	11.140	59¼	30¾			
11.160-80	11.171	77	13	11.167	66½	23½	11.170	59½	30½	11.167	54¼	35¾
11.180-00	11.194	77½	12½	11.187	66¾	23¼	11.198	59¾	30¼			
11.200-20	11.215	78	12	11.208	67	23				11.202	54½	35½
11.220-40	11.235	78½	11½	11.230	67¼	22¾	11.226	60	30	11.237	54¾	35¼
11.240-60	11.255	79	11	11.250	67½	22½	11.254	60¼	29¾			
11.260-80	11.273	79½	10½	11.270	67¾	22¼				11.271	55	35
11.280-00	11.292	80	10	11.290	68	22	11.282	60½	29½			
11.300-20				11.310	68¼	21¾	11.310	60¾	29¼	11.306	55¼	34¾
11.320-40				11.330	68½	21½						
11.340-60				11.350	68¾	21¼	11.338	61	29	11.340	55½	34½
11.360-80				11.369	69	21	11.366	61¼	28¾	11.374	55¾	34¼
11.380-00				11.388	69¼	20¾						

Approximate Lead	Gear on Worm 86, 1st on Stud 44, 2nd on Stud 48, Gear on Screw 64	Angle to Set Spiral Head (Degrees)	Angle to Set Vertical Att. (Degrees)	Gear on Worm 64, 1st on Stud 32, 2nd on Stud 56, Gear on Screw 72	Angle to Set Spiral Head (Degrees)	Angle to Set Vertical Att. (Degrees)	Gear on Worm 72, 1st on Stud 40, 2nd on Stud 44, Gear on Screw 48	Angle to Set Spiral Head (Degrees)	Angle to Set Vertical Att. (Degrees)	Gear on Worm 100, 1st on Stud 44, 2nd on Stud 56, Gear on Screw 72	Angle to Set Spiral Head (Degrees)	Angle to Set Vertical Att. (Degrees)
10.800-20	10.807	47½	42½	10.806	44	46						
10.820-40							10.825	41	49	10.822	37¾	52¼
10.840-60	10.850	47¾	42¼	10.854	44¼	45¾						
10.860-80												
10.880-00	10.893	48	42				10.880	41¼	48¾	10.883	38	52
10.900-20				10.903	44½	45½						
10.920-40	10.937	48¼	41¾				10.934	41½	48½			
10.940-60				10.951	44¾	45¼				10.944	38¼	51¾
10.960-80	10.979	48½	41½									
10.980-00							10.988	41¾	48¼			
11.000-20				11.000	45	45				11.004	38½	51½
11.020-40	11.021	48¾	41¼									
11.040-60				11.047	45¼	44¾	11.040	42	48			
11.060-80	11.063	49	41							11.064	38¾	51¼
11.080-00				11.094	45½	44½	11.094	42¼	47¾			
11.100-20	11.104	49¼	40¾									
11.120-40										11.124	39	51
11.140-60	11.147	49½	40½	11.142	45¾	44¼	11.147	42½	47½			
11.160-80												
11.180-00	11.188	49¾	40¼	11.190	46	44				11.184	39¼	50¾
11.200-20							11.200	42¾	47¼			
11.220-40	11.230	50	40	11.237	46¼	43¾						
11.240-60							11.253	43	47	11.244	39½	50½
11.260-80	11.270	50¼	39¾									
11.280-00				11.283	46½	43½						
11.300-20	11.310	50½	39½				11.306	43¼	46¾	11.303	39¾	50¼
11.320-40				11.330	46¾	43¼						
11.340-60	11.351	50¾	39¼				11.359	43½	46½			
11.360-80				11.377	47	43				11.362	40	
11.380-00	11.392	51	39									

Approximate Lead	Gear on Worm 72, 1st on Stud 44, 2nd on Stud 64, Gear on Screw 86	Angle to Set Spiral Head (Degrees)	Angle to Set Vertical Att. (Degrees)	Gear on Worm 56, 1st on Stud 24, 2nd on Stud 40, Gear on Screw 72	Angle to Set Spiral Head (Degrees)	Angle to Set Vertical Att. (Degrees)	Gear on Worm 86, 1st on Stud 40, 2nd on Stud 64, Gear on Screw 100	Angle to Set Spiral Head (Degrees)	Angle to Set Vertical Att. (Degrees)	Gear on Worm 86, 1st on Stud 44, 2nd on Stud 48, Gear on Screw 64	Angle to Set Spiral Head (Degrees)	Angle to Set Vertical Att. (Degrees)	Gear on Worm 64, 1st on Stud 32, 2nd on Stud 56, Gear on Screw 72	Angle to Set Spiral Head (Degrees)	Angle to Set Vertical Att. (Degrees)	Gear on Worm 72, 1st on Stud 40, 2nd on Stud 44, Gear on Screw 48	Angle to Set Spiral Head (Degrees)	Angle to Set Vertical Att. (Degrees)	Gear on Worm 100, 1st on Stud 44, 2nd on Stud 56, Gear on Screw 72	Angle to Set Spiral Head (Degrees)	Angle to Set Vertical Att. (Degrees)	Gear on Worm 86, 1st on Stud 28, 2nd on Stud 44, Gear on Screw 72	Angle to Set Spiral Head (Degrees)	Angle to Set Vertical Att. (Degrees)
11.400-20	11.406	69½	20½	11.419	61¾	28¼	11.408	56	34							11.410	43¾	46¼						
11.420-40	11.424	69¾	20¼							11.432	51¼	38¾	11.423	47¼	42¾				11.421	40¼	49¾	11.426	37½	52½
11.440-60	11.443	70	20	11.446	62	28	11.441	56¼	33¾															
11.460-80	11.479	70½	19½	11.472	62¼	27¾	11.475	56½	33½	11.472	51½	38½	11.469	47½	42½	11.462	44	46						
11.480-00	11.496	70¾	19¼	11.499	62½	27½													11.480	40½	49½	11.491	37¾	52¼
11.500-20	11.513	71	19				11.508	56¾	33¼	11.512	51¾	38¼	11.515	47¾	42¼	11.514	44¼	45¾						
11.520-40	11.530	71¼	18¾	11.525	62¾	27¼													11.539	40¾	49¼			
11.540-60	11.547	71½	18½	11.550	63	27	11.540	57	33	11.551	52	38										11.557	38	52
11.560-80	11.564	71¾	18¼	11.575	63¼	26¾	11.573	57¼	32¾				11.560	48	42	11.565	44½	45½						
11.580-00	11.580	72	18							11.591	52¼	37¾							11.597	41	49			
11.600-20	11.612	72½	17½	11.600	63½	26½	11.605	57½	32½				11.606	48¼	41¾	11.617	44¾	45¼						
11.620-40	11.629	72¾	17¼	11.625	63¾	26¼	11.637	57¾	32¼	11.630	52½	37½										11.620	38¼	51¾
11.640-60	11.644	73	17	11.650	64	26							11.650	48½	41½				11.656	41¼	48¾			
11.660-80	11.677	73½	16½	11.675	64¼	25¾	11.669	58	32	11.669	52¾	37¼				11.667	45	45						
11.680-00	11.690	73¾	16¼										11.695	48¾	41¼							11.685	38½	51½
11.700-20	11.704	74	16	11.700	64½	25½	11.700	58¼	31¾	11.707	53	37				11.718	45¼	44¾	11.713	41½	48½			
11.720-40	11.734	74½	15½	11.724	64¾	25¼	11.732	58½	31½															
11.740-60	11.748	74¾	15¼	11.748	65	25				11.745	53¼	36¾	11.740	49	41							11.749	38¾	51¼
11.760-80	11.761	75	15	11.772	65¼	24¾	11.763	58¾	31¼							11.769	45½	44½	11.771	41¾	48¼			
11.780-00	11.788	75½	14½	11.796	65½	24½	11.794	59	31	11.783	53½	36½	11.783	49¼	40¾									
11.800-20	11.815	76	14	11.819	65¾	24¼										11.819	45¾	44¼				11.813	39	51
11.820-40	11.828	76¼	13¾				11.825	59¼	30¾	11.821	53¾	36¼	11.828	49½	40½				11.829	42	48			
11.840-60	11.840	76½	13½	11.842	66	24	11.856	59½	30½															
11.860-80	11.864	77	13	11.864	66¼	23¾				11.860	54	36	11.872	49¾	40¼	11.870	46	44				11.876	39¼	50¾
11.880-00	11.888	77½	12½	11.888	66½	23½	11.887	59¾	30¼	11.896	54¼	35¾							11.885	42¼	47¾			
11.900-20	11.900	77¾	12¼	11.910	66¾	23¼	11.917	60	30				11.917	50	40	11.919	46¼	43¾						
11.920-40	11.932	78½	11½	11.932	67	23				11.934	54½	35½												
11.940-60	11.953	79	11	11.955	67¼	22¾	11.946	60¼	29¾										11.942	42½	47½	11.940	39½	50½
11.960-80	11.973	79½	10½	11.977	67½	22½	11.975	60½	29½	11.971	54¾	35¼	11.960	50¼	39¾	11.969	46½	43½						
11.980-00	11.992	80	10	11.998	67¾	22¼													11.999	42¾	47¼			

LEADS FROM 12.000 TO 12.600

Approximate Lead	Gear on Worm / 1st on Stud / 2nd on Stud / Gear on Screw	Angle to Set Spiral Head	Angle to Set Vertical Att.	Gear on Worm / 1st on Stud / 2nd on Stud / Gear on Screw	Angle to Set Spiral Head	Angle to Set Vertical Att.	Gear on Worm / 1st on Stud / 2nd on Stud / Gear on Screw	Angle to Set Spiral Head	Angle to Set Vertical Att.	Gear on Worm / 1st on Stud / 2nd on Stud / Gear on Screw	Angle to Set Spiral Head	Angle to Set Vertical Att.	Gear on Worm / 1st on Stud / 2nd on Stud / Gear on Screw	Angle to Set Spiral Head	Angle to Set Vertical Att.	Gear on Worm / 1st on Stud / 2nd on Stud / Gear on Screw	Angle to Set Spiral Head	Angle to Set Vertical Att.	Gear on Worm / 1st on Stud / 2nd on Stud / Gear on Screw	Angle to Set Spiral Head	Angle to Set Vertical Att.	Gear on Worm / 1st on Stud / 2nd on Stud / Gear on Screw	Angle to Set Spiral Head	Angle to Set Vertical Att.
	56 / 24 / 40 / 72	Degrees	Degrees	86 / 40 / 64 / 100	Degrees	Degrees	86 / 44 / 48 / 64	Degrees	Degrees	64 / 32 / 56 / 72	Degrees	Degrees	72 / 40 / 44 / 48	Degrees	Degrees	100 / 44 / 56 / 72	Degrees	Degrees	86 / 28 / 44 / 72	Degrees	Degrees	100 / 28 / 48 / 86	Degrees	Degrees
12.000-20	12.019	68	22	12.006	60¾	29¼	12.008	55	35	12.003	50½	39½	12.019	46¾	43¼				12.003	39¾	50¼			
12.020-40				12.035	61	29																		
12.040-60	12.040	68¼	21¾				12.044	55¼	34¾	12.046	50¾	39¼				12.056	43	47						
12.060-80	12.061	68½	21½	12.064	61¼	28¾							12.068	47	43				12.066	40	50	12.065	37¼	52¾
12.080-00	12.082	68¾	21¼	12.093	61½	28½	12.080	55½	34½	12.090	51	39												
12.100-20	12.102	69	21				12.117	55¾	34¼				12.117	47¼	42¾	12.112	43¼	46¾						
12.120-40	12.122	69¼	20¾	12.121	61¾	28¼				12.132	51¼	38¾							12.128	40¼	49¾	12.134	37½	52½
12.140-60	12.142	69½	20½	12.150	62	28	12.152	56	34															
12.160-80	12.162	69¾	20¼	12.178	62¼	27¾				12.174	51½	38½	12.165	47½	42½	12.169	43½	46½						
12.180-00	12.181	70	20				12.189	56¼	33¾										12.190	40½	49½			
12.200-20	12.201	70¼	19¾	12.206	62½	27½				12.216	51¾	38¼	12.214	47¾	42¼							12.203	37¾	52¼
12.220-40	12.238	70¾	19¼	12.233	62¾	27¼	12.224	56½	33½							12.223	43¾	46¼						
12.240-60	12.256	71	19							12.258	52	38							12.252	40¾	49¼			
12.260-80	12.275	71¼	18¾	12.260	63	27	12.260	56¾	33¼				12.262	48	42							12.272	38	52
12.280-00	12.292	71½	18½	12.287	63¼	26¾	12.294	57	33							12.280	44	46						
12.300-20	12.310	71¾	18¼	12.314	63½	26½				12.300	52¼	37¾	12.310	48¼	41¾				12.315	41	49			
12.320-40	12.328	72	18				12.330	57¼	32¾							12.335	44¼	45¾						
12.340-60	12.346	72¼	17¾	12.341	63¾	26¼				12.341	52½	37½	12.358	48½	41½							12.340	38¼	51¾
12.360-80	12.362	72½	17½	12.366	64	26	12.363	57½	32½										12.376	41¼	48¾			
12.380-00	12.395	73	17	12.393	64¼	25¾	12.397	57¾	32¼	12.382	52¾	37¼				12.390	44½	45½						
12.400-20	12.412	73¼	16¾	12.419	64½	25½							12.406	48¾	41¼							12.409	38½	51½
12.420-40	12.428	73½	16½				12.430	58	32	12.422	53	37							12.439	41½	48½			
12.440-60	12.444	73¾	16¼	12.445	64¾	25¼										12.444	44¾	45¼						
12.460-80	12.460	74	16	12.470	65	25	12.464	58¼	31¾	12.463	53¼	36¾	12.453	49	41									
12.480-00	12.491	74½	15½	12.496	65¼	24¾	12.498	58½	31½													12.477	38¾	51¼
12.500-20	12.506	74¾	15¼							12.504	53½	36½	12.500	49¼	40¾	12.500	45	45	12.500	41¾	48¼			
12.520-40	12.521	75	15	12.520	65½	24½	12.531	58¾	31¼															
12.540-60	12.550	75½	14½	12.545	65¾	24¼				12.544	53¾	36¼	12.547	49½	40½	12.553	45¼	44¾				12.545	39	51
12.560-80	12.578	76	14	12.570	66	24	12.564	59	31										12.560	42	48			
12.580-00	12.591	76¼	13¾	12.594	66¼	23¾				12.585	54	36	12.593	49¾	40¼									

Approximate Lead	Gear on Worm 86, 1st on Stud 40, 2nd on Stud 64, Gear on Screw 100	Angle to Set Spiral Head (Degrees)	Angle to Set Vertical Att. (Degrees)	Gear on Worm 86, 1st on Stud 44, 2nd on Stud 48, Gear on Screw 64	Angle to Set Spiral Head (Degrees)	Angle to Set Vertical Att. (Degrees)	Gear on Worm 64, 1st on Stud 32, 2nd on Stud 56, Gear on Screw 72	Angle to Set Spiral Head (Degrees)	Angle to Set Vertical Att. (Degrees)	Gear on Worm 72, 1st on Stud 40, 2nd on Stud 44, Gear on Screw 48	Angle to Set Spiral Head (Degrees)	Angle to Set Vertical Att. (Degrees)	Gear on Worm 100, 1st on Stud 44, 2nd on Stud 56, Gear on Screw 72	Angle to Set Spiral Head (Degrees)	Angle to Set Vertical Att. (Degrees)	Gear on Worm 86, 1st on Stud 28, 2nd on Stud 44, Gear on Screw 72	Angle to Set Spiral Head (Degrees)	Angle to Set Vertical Att. (Degrees)	Gear on Worm 100, 1st on Stud 28, 2nd on Stud 48, Gear on Screw 86	Angle to Set Spiral Head (Degrees)	Angle to Set Vertical Att. (Degrees)	Gear on Worm 86, 1st on Stud 32, 2nd on Stud 44, Gear on Screw 56	Angle to Set Spiral Head (Degrees)	Angle to Set Vertical Att. (Degrees)
12.600-20	12.619	66½	23½										12.608	45½	44½				12.613	39¼	50¾			
12.620-40				12.630	59½	30½	12.625	54¼	35¾							12.621	42¼	47¾				12.633	36¾	53¼
12.640-60	12.643	66¾	23¼							12.640	50	40												
12.660-80	12.666	67	23	12.662	59¾	30¼	12.664	54½	35½				12.661	45¾	44¼									
12.680-00	12.690	67¼	22¾	12.695	60	30				12.686	50¼	39¾				12.681	42½	47½	12.680	39½	50½			
12.700-20	12.713	67½	22½				12.703	54¾	35¼				12.716	46	44							12.709	37	53
12.720-40	12.736	67¾	22¼	12.726	60¼	29¾				12.732	50½	39½												
12.740-60	12.758	68	22	12.758	60½	29½	12.742	55	35							12.741	42¾	47¼	12.747	39¾	50¼			
12.760-80										12.778	50¾	39¼	12.769	46¼	43¾									
12.780-00	12.780	68¼	21¾	12.790	60¾	29¼	12.781	55¼	34¾													12.782	37¼	52¾
12.800-20	12.802	68½	21½													12.802	43	47	12.813	40	50			
12.820-40	12.824	68¾	21¼	12.820	61	29	12.820	55½	34½	12.823	51	39	12.822	46½	43½									
12.840-60	12.846	69	21	12.852	61¼	28¾	12.858	55¾	34¼													12.854	37½	52½
12.860-80	12.868	69¼	20¾							12.868	51¼	38¾	12.875	46¾	43¼	12.861	43¼	46¾	12.879	40¼	49¾			
12.880-00	12.889	69½	20½	12.882	61½	28½	12.896	56	34															
12.900-20	12.910	69¾	20¼	12.912	61¾	28¼				12.913	51½	38½												
12.920-40	12.930	70	20				12.933	56¼	33¾				12.929	47	43	12.921	43½	46½				12.928	37¾	52¼
12.940-60	12.950	70¼	19¾	12.943	62	28				12.958	51¾	38¼							12.945	40½	49½			
12.960-80	12.971	70½	19½	12.973	62¼	27¾	12.972	56½	33½															
12.980-00	12.990	70¾	19¼										12.980	47¼	42¾	12.980	43¾	46¼						
13.000-20	13.010	71	19	13.002	62½	27½	13.010	56¾	33¼	13.003	52	38							13.011	40¾	49¼	13.000	38	52
13.020-40	13.030	71¼	18¾	13.032	62¾	27¼							13.033	47½	42½									
13.040-60	13.049	71½	18½				13.046	57	33	13.048	52¼	37¾				13.040	44	46						
13.060-80	13.068	71¾	18¼	13.061	63	27													13.078	41	49	13.073	38¼	51¾
13.080-00	13.086	72	18	13.090	63¼	26¾	13.083	57¼	32¾	13.090	52½	37½	13.085	47¾	42¼	13.098	44¼	45¾						
13.100-20	13.104	72¼	17¾	13.118	63½	26½	13.119	57½	32½															
13.120-40	13.122	72½	17½							13.134	52¾	37¼	13.137	48	42									
13.140-60	13.158	73	17	13.147	63¾	26¼	13.155	57¾	32¼							13.157	44½	45½	13.143	41¼	48¾	13.145	38½	51½
13.160-80	13.175	73¼	16¾	13.174	64	26				13.177	53	37												
13.180-00	13.192	73½	16½				13.190	58	32				13.188	48¼	41¾									

LEADS FROM 13.200 TO 13.800

Approximate Lead	Gear on Worm 86, 1st on Stud 44, 2nd on Stud 48, Gear on Screw 64	Angle to Set Spiral Head (Degrees)	Angle to Set Vertical Att. (Degrees)	Gear on Worm 64, 1st on Stud 32, 2nd on Stud 56, Gear on Screw 72	Angle to Set Spiral Head (Degrees)	Angle to Set Vertical Att. (Degrees)	Gear on Worm 72, 1st on Stud 40, 2nd on Stud 44, Gear on Screw 48	Angle to Set Spiral Head (Degrees)	Angle to Set Vertical Att. (Degrees)	Gear on Worm 100, 1st on Stud 44, 2nd on Stud 56, Gear on Screw 72	Angle to Set Spiral Head (Degrees)	Angle to Set Vertical Att. (Degrees)	Gear on Worm 86, 1st on Stud 28, 2nd on Stud 44, Gear on Screw 72	Angle to Set Spiral Head (Degrees)	Angle to Set Vertical Att. (Degrees)	Gear on Worm 100, 1st on Stud 28, 2nd on Stud 48, Gear on Screw 86	Angle to Set Spiral Head (Degrees)	Angle to Set Vertical Att. (Degrees)	Gear on Worm 86, 1st on Stud 32, 2nd on Stud 44, Gear on Screw 56	Angle to Set Spiral Head (Degrees)	Angle to Set Vertical Att. (Degrees)	Gear on Worm 72, 1st on Stud 28, 2nd on Stud 56, Gear on Screw 64	Angle to Set Spiral Head (Degrees)	Angle to Set Vertical Att. (Degrees)
13.200-20	13.202	64¼	25¾										13.215	44¾	45¼	13.209	41½	48½	13.217	38¾	51¼			
13.220-40	13.230	64½	25½	13.227	58¼	31¾	13.220	53¼	36¾	13.239	48½	41½										13.225	36	54
13.240-60	13.258	64¾	25¼																					
13.260-80				13.262	58½	31½	13.263	53½	36½				13.273	45	45	13.274	41¾	48¼						
13.280-00	13.285	65	25	13.298	58¾	31¼				13.290	48¾	41¼							13.289	39	51			
13.300-20	13.311	65¼	24¾				13.306	53¾	36¼													13.304	36¼	53¾
13.320-40	13.339	65½	24½	13.333	59	31							13.330	45¼	44¾									
13.340-60							13.349	54	36	13.340	49	41				13.340	42	48						
13.360-80	13.364	65¾	24¼	13.368	59¼	30¾													13.360	39¼	50¾			
13.380-00	13.390	66	24				13.392	54¼	35¾	13.391	49¼	40¾	13.388	45½	44½							13.384	36½	53½
13.400-20	13.417	66¼	23¾	13.403	59½	30½										13.403	42¼	47¾						
13.420-40				13.438	59¾	30¼	13.433	54½	35½										13.432	39½	50½			
13.440-60	13.442	66½	23½							13.441	49½	40½	13.445	45¾	44¼									
13.460-80	13.467	66¾	23¼	13.471	60	30	13.475	54¾	35¼							13.467	42½	47½				13.462	36¾	53¼
13.480-00	13.493	67	23							13.491	49¾	40¼												
13.500-20	13.519	67¼	22¾	13.505	60¼	29¾	13.517	55	35				13.502	46	44				13.503	39¾	50¼			
13.520-40				13.538	60½	29½										13.530	42¾	47¼						
13.540-60	13.543	67½	22½				13.558	55¼	34¾	13.541	50	40										13.541	37	53
13.560-80	13.568	67¾	22¼	13.572	60¾	29¼							13.560	46¼	43¾				13.573	40	50			
13.580-00	13.590	68	22				13.599	55½	34½	13.590	50¼	39¾				13.595	43	47						
13.600-20	13.615	68¼	21¾	13.605	61	29							13.616	46½	43½							13.618	37¼	52¾
13.620-40	13.639	68½	21½	13.638	61¼	28¾																		
13.640-60							13.640	55¾	34¼	13.640	50½	39½				13.659	43¼	46¾	13.643	40¼	49¾			
13.660-80	13.662	68¾	21¼	13.670	61½	28½							13.672	46¾	43¼									
13.680-00	13.685	69	21				13.680	56	34	13.689	50¾	39¼										13.696	37½	52½
13.700-20	13.709	69¼	20¾	13.702	61¾	28¼													13.714	40½	49½			
13.720-40	13.730	69½	20½	13.735	62	28	13.720	56¼	33¾	13.738	51	39	13.723	47	43	13.722	43½	46½						
13.740-60	13.752	69¾	20¼																					
13.760-80	13.775	70	20	13.767	62¼	27¾	13.760	56½	33½													13.775	37¾	52¼
13.780-00	13.797	70¼	19¾	13.798	62½	27½				13.786	51¼	38¾	13.784	47¼	42¾	13.784	43¾	46¼	13.783	40¾	49¼			

Approximate Lead	Gear on Worm 86, 1st on Stud 44, 2nd on Stud 48, Gear on Screw 64	Angle to Set Spiral Head (Degrees)	Angle to Set Vertical Att. (Degrees)	Gear on Worm 64, 1st on Stud 32, 2nd on Stud 56, Gear on Screw 72	Angle to Set Spiral Head (Degrees)	Angle to Set Vertical Att. (Degrees)	Gear on Worm 72, 1st on Stud 40, 2nd on Stud 44, Gear on Screw 48	Angle to Set Spiral Head (Degrees)	Angle to Set Vertical Att. (Degrees)	Gear on Worm 100, 1st on Stud 44, 2nd on Stud 56, Gear on Screw 72	Angle to Set Spiral Head (Degrees)	Angle to Set Vertical Att. (Degrees)
13.800-20	13.818	70½	19½				13.800	56¾	33¼			
13.820-40	13.839	70¾	19¼	13.830	62¾	27¼	13.839	57	33	13.833	51½	38½
13.840-60	13.859	71	19									
13.860-80				13.860	63	27	13.878	57¼	32¾			
13.880-00	13.880	71¼	18¾	13.890	63¼	26¾	13.916	57½	32½	13.882	51¾	38¼
13.900-20	13.900	71½	18½									
13.920-40	13.920	71¾	18¼	13.920	63½	26½				13.930	52	38
13.940-60	13.940	72	18	13.950	63¾	26¼	13.954	57¾	32¼			
13.960-80	13.960	72¼	17¾							13.978	52¼	37¾
13.980-00	13.998	72¾	17¼	13.980	64	26	13.992	58	32			
14.000-20	14.017	73	17	14.010	64¼	25¾						
14.020-40	14.035	73¼	16¾				14.030	58¼	31¾	14.024	52½	37½
14.040-60	14.053	73½	16½	14.040	64½	25½						
14.060-80	14.071	73¾	16¼	14.069	64¾	25¼	14.069	58½	31½	14.070	52¾	37¼
14.080-00	14.090	74	16	14.098	65	25						
14.100-20	14.108	74¼	15¾				14.106	58¾	31¼	14.118	53	37
14.120-40	14.125	74½	15½	14.127	65¼	24¾						
14.140-60	14.159	75	15	14.155	65½	24½	14.143	59	31			
14.160-80	14.175	75¼	14¾							14.163	53¼	36¾
14.180-00	14.191	75½	14½	14.183	65¾	24¼	14.180	59¼	30¾			
14.200-20	14.207	75¾	14¼	14.211	66	24	14.217	59½	30½	14.210	53½	36½
14.220-40	14.222	76	14	14.238	66¼	23¾						
14.240-60	14.253	76½	13½				14.253	59¾	30¼	14.255	53¾	36¼
14.260-80	14.268	76¾	13¼	14.265	66½	23½						
14.280-00	14.283	77	13	14.291	66¾	23¼	14.290	60	30			
14.300-20	14.310	77½	12½	14.318	67	23				14.300	54	36
14.320-40	14.338	78	12				14.325	60¼	29¾			
14.340-60	14.351	78¼	11¾	14.345	67¼	22¾				14.345	54¼	35¾
14.360-80	14.364	78½	11½	14.371	67½	22½	14.360	60½	29½			
14.380-00	14.390	79	11	14.398	67¾	22¼				14.390	54½	35½

Approximate Lead	Gear on Worm 86, 1st on Stud 28, 2nd on Stud 44, Gear on Screw 72	Angle to Set Spiral Head (Degrees)	Angle to Set Vertical Att. (Degrees)	Gear on Worm 100, 1st on Stud 28, 2nd on Stud 48, Gear on Screw 86	Angle to Set Spiral Head (Degrees)	Angle to Set Vertical Att. (Degrees)	Gear on Worm 86, 1st on Stud 32, 2nd on Stud 44, Gear on Screw 56	Angle to Set Spiral Head (Degrees)	Angle to Set Vertical Att. (Degrees)	Gear on Worm 72, 1st on Stud 28, 2nd on Stud 56, Gear on Screw 64	Angle to Set Spiral Head (Degrees)	Angle to Set Vertical Att. (Degrees)
13.800-20												
13.820-40												
13.840-60	13.840	47½	42½	13.847	44	46						
13.860-80							13.854	41	49	13.853	38	52
13.880-00	13.895	47¾	42¼									
13.900-20				13.909	44¼	45¾						
13.920-40							13.923	41¼	48¾	13.929	38¼	51¾
13.940-60	13.950	48	42									
13.960-80				13.971	44½	45½						
13.980-00							13.992	41½	48½			
14.000-20	14.004	48¼	41¾							14.007	38½	51½
14.020-40				14.033	44¾	45¼						
14.040-60	14.058	48½	41½									
14.060-80							14.061	41¾	48¼			
14.080-00				14.095	45	45				14.083	38¾	51¼
14.100-20	14.113	48¾	41¼									
14.120-40							14.129	42	48			
14.140-60				14.157	45¼	44¾						
14.160-80	14.166	49	41							14.160	39	51
14.180-00							14.198	42¼	47¾			
14.200-20				14.218	45½	44½						
14.220-40	14.220	49¼	40¾							14.236	39¼	50¾
14.240-60												
14.260-80	14.272	49½	40½	14.279	45¾	44¼	14.266	42½	47½			
14.280-00												
14.300-20										14.313	39½	50½
14.320-40	14.326	49¾	40¼				14.333	42¾	47¼			
14.340-60				14.340	46	44						
14.360-80												
14.380-00	14.380	50	40							14.388	39¾	50¼

LEADS FROM 14.400 TO 15.000

Approximate Lead	Gear on Worm 64, 1st on Stud 32, 2nd on Stud 56, Gear on Screw 72	Angle to Set Spiral Head (Degrees)	Angle to Set Vertical Att. (Degrees)	Gear on Worm 72, 1st on Stud 40, 2nd on Stud 44, Gear on Screw 48	Angle to Set Spiral Head (Degrees)	Angle to Set Vertical Att. (Degrees)	Gear on Worm 100, 1st on Stud 44, 2nd on Stud 56, Gear on Screw 72	Angle to Set Spiral Head (Degrees)	Angle to Set Vertical Att. (Degrees)	Gear on Worm 86, 1st on Stud 28, 2nd on Stud 44, Gear on Screw 72	Angle to Set Spiral Head (Degrees)	Angle to Set Vertical Att. (Degrees)
14.400-20												
14.420-40	14.422	68	22	14.431	61	29	14.437	54¾	35¼	14.431	50¼	39¾
14.440-60	14.448	68¼	21¾									
14.460-80	14.473	68½	21½	14.466	61¼	28¾						
14.480-00	14.498	68¾	21¼				14.480	55	35	14.483	50½	39½
14.500-20				14.500	61½	28½						
14.520-40	14.522	69	21	14.534	61¾	28¼	14.525	55¼	34¾	14.536	50¾	39¼
14.540-60	14.547	69¼	20¾									
14.560-80	14.570	69½	20½	14.568	62	28	14.569	55½	34½			
14.580-00	14.594	69¾	20¼							14.588	51	39
14.600-20	14.618	70	20	14.602	62¼	27¾	14.611	55¾	34¼			
14.620-40				14.636	62½	27½						
14.640-60	14.640	70¼	19¾				14.654	56	34	14.640	51¼	38¾
14.660-80	14.663	70½	19½	14.670	62¾	27¼						
14.680-00	14.686	70¾	19¼				14.698	56¼	33¼	14.690	51½	38½
14.700-20	14.706	71	19	14.702	63	27						
14.720-40	14.730	71¼	18¾	14.733	63¼	26¾						
14.740-60	14.750	71½	18½				14.740	56½	33½	14.740	51¾	38¼
14.760-80	14.771	71¾	18¼	14.766	63½	26½						
14.780-00	14.793	72	18	14.790	63¾	26¼	14.783	56¾	33¼	14.792	52	38
14.800-20	14.814	72¼	17¾									
14.820-40	14.834	72½	17½	14.829	64	26	14.825	57	33			
14.840-60	14.854	72¾	17¼							14.842	52¼	37¾
14.860-80	14.874	73	17	14.861	64¼	25¾	14.868	57¼	32¾			
14.880-00	14.894	73¼	16¾	14.892	64½	25½				14.892	52½	37½
14.900-20	14.913	73½	16½				14.908	57½	32½			
14.920-40	14.932	73¾	16¼	14.923	64¾	25¼						
14.940-60	14.952	74	16	14.953	65	25	14.950	57¾	32¼	14.940	52¾	37¼
14.960-80	14.970	74¼	15¾									
14.980-00	14.990	74½	15½	14.984	65¼	24¾	14.990	58	32	14.990	53	37

Approximate Lead	Gear on Worm 100, 1st on Stud 28, 2nd on Stud 48, Gear on Screw 86	Angle to Set Spiral Head (Degrees)	Angle to Set Vertical Att. (Degrees)	Gear on Worm 86, 1st on Stud 32, 2nd on Stud 44, Gear on Screw 56	Angle to Set Spiral Head (Degrees)	Angle to Set Vertical Att. (Degrees)	Gear on Worm 72, 1st on Stud 28, 2nd on Stud 56, Gear on Screw 64	Angle to Set Spiral Head (Degrees)	Angle to Set Vertical Att. (Degrees)	Gear on Worm 64, 1st on Stud 40, 2nd on Stud 72, Gear on Screw 48	Angle to Set Spiral Head (Degrees)	Angle to Set Vertical Att. (Degrees)
14.400-20	14.400	46¼	43¾	14.401	43	47						
14.420-40												
14.440-60										14.444	37	53
14.460-80	14.460	46½	43½	14.469	43¼	46¾	14.462	40	50			
14.480-00												
14.500-20	14.519	46¾	43¼									
14.520-40				14.536	43½	46½	14.538	40¼	49¾	14.526	37¼	52¾
14.540-60												
14.560-80	14.579	47	43									
14.580-00												
14.600-20				14.602	43¾	46¼	14.612	40½	49½	14.610	37½	52½
14.620-40	14.638	47¼	42¾									
14.640-60												
14.660-80				14.669	44	46						
14.680-00	14.697	47½	42½				14.687	40¾	49¼	14.693	37¾	52¼
14.700-20												
14.720-40				14.734	44¼	45¾						
14.740-60	14.755	47¾	42¼									
14.760-80							14.761	41	49	14.775	38	52
14.780-00												
14.800-20	14.813	48	42	14.800	44½	45½						
14.820-40							14.835	41¼	48¾			
14.840-60										14.858	38¼	51¾
14.860-80	14.872	48¼	41¾	14.867	44¾	45¼						
14.880-00												
14.900-20							14.910	41½	48½			
14.920-40	14.930	48½	41½	14.931	45	45						
14.940-60										14.940	38½	51½
14.960-80												
14.980-00	14.988	48¾	41¼	14.996	45¼	44¾	14.983	41¾	48¼			

Approximate Lead	Gear on Worm 72, 1st on Stud 40, 2nd on Stud 44, Gear on Screw 48	Angle to Set Spiral Head (Degrees)	Angle to Set Vertical Att. (Degrees)	Gear on Worm 100, 1st on Stud 44, 2nd on Stud 56, Gear on Screw 72	Angle to Set Spiral Head (Degrees)	Angle to Set Vertical Att. (Degrees)	Gear on Worm 86, 1st on Stud 28, 2nd on Stud 44, Gear on Screw 72	Angle to Set Spiral Head (Degrees)	Angle to Set Vertical Att. (Degrees)	Gear on Worm 100, 1st on Stud 28, 2nd on Stud 48, Gear on Screw 86	Angle to Set Spiral Head (Degrees)	Angle to Set Vertical Att. (Degrees)	Gear on Worm 86, 1st on Stud 32, 2nd on Stud 44, Gear on Screw 56	Angle to Set Spiral Head (Degrees)	Angle to Set Vertical Att. (Degrees)	Gear on Worm 72, 1st on Stud 28, 2nd on Stud 56, Gear on Screw 64	Angle to Set Spiral Head (Degrees)	Angle to Set Vertical Att. (Degrees)	Gear on Worm 64, 1st on Stud 40, 2nd on Stud 72, Gear on Screw 48	Angle to Set Spiral Head (Degrees)	Angle to Set Vertical Att. (Degrees)	Gear on Worm 100, 1st on Stud 64, 2nd on Stud 72, Gear on Screw 44	Angle to Set Spiral Head (Degrees)	Angle to Set Vertical Att. (Degrees)
15.000-40	15.014	65½	24½	15.030	58¼	31¾													15.022	38¾	51¼	15.028	36	54
15.040-80	15.073	66	24	15.071	58½	31½	15.040	53¼	36¾	15.043	49	41	15.060	45½	44½	15.055	42	48						
15.080-20	15.101	66¼	23¾	15.111	58¾	31¼	15.089	53½	36½	15.100	49¼	40¾							15.105	39	51	15.118	36¼	53¾
15.120-60	15.131	66½	23½	15.151	59	31	15.138	53¾	36¼	15.158	49½	40½	15.125	45¾	44¼	15.128	42¼	47¾						
15.160-00	15.160	66¾	23¼	15.191	59¼	30¾	15.186	54	36				15.190	46	44				15.185	39¼	50¾			
15.200-40	15.217	67¼	22¾	15.230	59½	30½	15.233	54¼	35¾	15.213	49¾	40¼				15.200	42½	47½				15.208	36½	53½
15.240-80	15.244	67½	22½	15.270	59¾	30¼				15.270	50	40	15.253	46¼	43¾	15.272	42¾	47¼	15.268	39½	50½			
15.280-20	15.299	68	22	15.309	60	30	15.281	54½	35½				15.317	46½	43½							15.298	36¾	53¼
15.320-60	15.352	68½	21½	15.347	60¼	29¾	15.329	54¾	35¼	15.325	50¼	39¾				15.345	43	47	15.347	39¾	50¼			
15.360-00	15.379	68¾	21¼	15.385	60½	29½	15.376	55	35	15.381	50½	39½	15.380	46¾	43¼							15.388	37	53
15.400-40	15.404	69	21	15.422	60¾	29¼	15.423	55¼	34¾	15.426	50¾	39¼				15.417	43¼	46¾	15.427	40	50			
15.440-80	15.455	69½	20½	15.460	61	29	15.470	55½	34½				15.443	47	43							15.476	37¼	52¾
15.480-20	15.505	70	20	15.498	61¼	28¾	15.515	55¾	34¼	15.491	51	39	15.507	47¼	42¾	15.489	43½	46½	15.507	40¼	49¾			
15.520-60	15.554	70½	19½	15.535	61½	28½				15.546	51¼	38¾				15.559	43¾	46¼						
15.560-00	15.577	70¾	19¼	15.572	61¾	28¼	15.561	56	34				15.569	47½	42½				15.587	40½	49½	15.564	37½	52½
15.600-40	15.624	71¼	18¾	15.609	62	28	15.607	56¼	33¾	15.600	51½	38½	15.630	47¾	42¼	15.630	44	46						
15.640-80	15.647	71½	18½	15.643	62¼	27¾	15.653	56½	33½	15.653	51¾	38¼							15.667	40¾	49¼	15.653	37¾	52¼
15.680-20	15.714	72¼	17¾	15.680	62½	27½	15.693	56¾	33¼	15.709	52	38	15.692	48	42	15.700	44¼	45¾						
15.720-60	15.735	72½	17½	15.750	63	27	15.742	57	33				15.754	48¼	41¾				15.745	41	49	15.741	38	52
15.760-00	15.798	73¼	16¾	15.783	63¼	26¾	15.788	57¼	32¾	15.761	52¼	37¾				15.770	44½	45½						
15.800-40	15.819	73½	16½	15.819	63½	26½	15.830	57½	32½	15.814	52½	37½	15.815	48½	41½				15.824	41¼	48¾	15.829	38¼	51¾
15.840-80	15.860	74	16	15.853	63¾	26¼	15.874	57¾	32¼	15.868	52¾	37¼	15.876	48¾	41¼	15.840	44¾	45¼						
15.880-20	15.900	74½	15½	15.887	64	26	15.918	58	32	15.919	53	37				15.910	45	45	15.903	41½	48½	15.916	38½	51½
15.920-60	15.938	75	15	15.954	64½	25½							15.936	49	41									
15.960-00	15.973	75½	14½	15.988	64¾	25¼	15.960	58¼	31¾	15.971	53¼	36¾	15.996	49¼	40¾	15.980	45¼	44¾	15.981	41¾	48¼			
16.000-40	16.010	76	14	16.020	65	25	16.004	58½	31½	16.023	53½	36½										16.002	38¾	51¼
16.040-80	16.043	76½	13½	16.053	65¼	24¾	16.047	58¾	31¼	16.075	53¾	36¼	16.057	49½	40½	16.048	45½	44½	16.060	42	48			
16.080-20	16.109	77½	12½	16.085	65½	24½	16.089	59	31				16.116	49¾	40¼	16.117	45¾	44¼				16.090	39	51
16.120-60	16.140	78	12	16.148	66	24	16.131	59¼	30¾	16.127	54	36							16.137	42¼	47¾			
16.160-00	16.197	79	11	16.180	66¼	23¾	16.173	59½	30½	16.177	54¼	35¾	16.176	50	40	16.185	46	44				16.177	39¼	50¾

LEADS FROM 16.200 TO 17.400

Approximate Lead	Gear on Worm 100, 1st on Stud 44, 2nd on Stud 56, Gear on Screw 72	Angle to Set Spiral Head (Degrees)	Angle to Set Vertical Att. (Degrees)	Gear on Worm 86, 1st on Stud 28, 2nd on Stud 44, Gear on Screw 72	Angle to Set Spiral Head (Degrees)	Angle to Set Vertical Att. (Degrees)	Gear on Worm 100, 1st on Stud 28, 2nd on Stud 48, Gear on Screw 86	Angle to Set Spiral Head (Degrees)	Angle to Set Vertical Att. (Degrees)	Gear on Worm 86, 1st on Stud 32, 2nd on Stud 44, Gear on Screw 56	Angle to Set Spiral Head (Degrees)	Angle to Set Vertical Att. (Degrees)	Gear on Worm 72, 1st on Stud 28, 2nd on Stud 56, Gear on Screw 64	Angle to Set Spiral Head (Degrees)	Angle to Set Vertical Att. (Degrees)	Gear on Worm 64, 1st on Stud 40, 2nd on Stud 72, Gear on Screw 48	Angle to Set Spiral Head (Degrees)	Angle to Set Vertical Att. (Degrees)	Gear on Worm 100, 1st on Stud 64, 2nd on Stud 72, Gear on Screw 44	Angle to Set Spiral Head (Degrees)	Angle to Set Vertical Att. (Degrees)	Gear on Worm 100, 1st on Stud 72, 2nd on Stud 86, Gear on Screw 44	Angle to Set Spiral Head (Degrees)	Angle to Set Vertical Att. (Degrees)
16.200-40	16.210	66½	23½	16.214	59¾	30¼	16.228	54½	35½	16.235	50¼	39¾				16.213	42½	47½						
16.240-80	16.271	67	23	16.256	60	30	16.279	54¾	35¼				16.253	46¼	43¾				16.263	39½	50½	16.242	36¾	53¼
16.280-20	16.301	67¼	22¾	16.296	60¼	29¾				16.294	50½	39½				16.290	42¾	47¼						
16.320-60	16.331	67½	22½	16.337	60½	29½	16.329	55	35	16.352	50¾	39¼	16.321	46½	43½				16.349	39¾	50¾	16.337	37	53
16.360-00	16.389	68	22	16.378	60¾	29¼	16.379	55¼	34¾				16.389	46¾	43¼	16.368	43	47						
16.400-40	16.419	68¼	21¾	16.417	61	29	16.429	55½	34½	16.410	51	39							16.434	40	50	16.432	37¼	52¾
16.440-80	16.447	68½	21½	16.457	61¼	28¾	16.477	55¾	34¼	16.468	51¼	38¾	16.455	47	43	16.445	43¼	46¾						
16.480-20	16.502	69	21	16.497	61½	28½																		
16.520-60	16.558	69½	20½	16.535	61¾	28¼	16.526	56	34	16.525	51½	38½	16.522	47¼	42¾	16.520	43½	46½	16.520	40¼	49¾	16.525	37½	52½
16.560-00	16.584	69¾	20¼	16.574	62	28	16.574	56¼	33¾	16.582	51¾	38¼	16.589	47½	42½	16.596	43¾	46¼						
16.600-40	16.610	70	20	16.612	62¼	27¾	16.622	56½	33½										16.604	40½	49½	16.620	37¾	52¼
16.640-80	16.663	70½	19½	16.650	62½	27½	16.670	56¾	33¼	16.640	52	38	16.655	47¾	42¼	16.671	44	46						
16.680-20	16.711	71	19	16.688	62¾	27¼	16.718	57	33	16.698	52¼	37¾							16.689	40¾	49¼	16.713	38	52
16.720-60	16.738	71¼	18¾	16.725	63	27				16.752	52½	37½	16.720	48	42	16.748	44¼	45¾						
16.760-00	16.762	71½	18½	16.798	63½	26½	16.766	57¼	32¾				16.787	48¼	41¾				16.773	41	49			
16.800-40	16.810	72	18	16.835	63¾	26¼	16.811	57½	32½	16.808	52¾	37¼				16.822	44½	45½				16.805	38¼	51¾
16.840-80	16.858	72½	17½	16.870	64	26	16.858	57¾	32¼	16.863	53	37	16.851	48½	41½				16.858	41¼	48¾			
16.880-20	16.881	72¾	17¼	16.906	64¼	25¾	16.904	58	32	16.919	53¼	36¾	16.918	48¾	41¼	16.898	44¾	45¼				16.900	38½	51½
16.920-60	16.925	73¼	16¾	16.941	64½	25½	16.950	58¼	31¾										16.942	41½	48½			
16.960-00	16.991	74	16	16.977	64¾	25¼	16.996	58½	31½	16.974	53½	36½	16.980	49	41	16.971	45	45				16.990	38¾	51¼
17.000-40	17.033	74½	15½	17.011	65	25				17.030	53¾	36¼							17.025	41¾	48¼			
17.040-80	17.074	75	15	17.047	65¼	24¾	17.040	58¾	31¼				17.045	49¼	40¾	17.045	45¼	44¾						
17.080-20	17.093	75¼	14¾	17.080	65½	24½	17.086	59	31	17.083	54	36	17.110	49½	40½	17.119	45½	44½	17.109	42	48	17.084	39	51
17.120-60	17.151	76	14	17.148	66	24	17.130	59¼	30¾	17.137	54¼	35¾												
17.160-00	17.188	76½	13½	17.180	66¼	23¾	17.175	59½	30½	17.190	54½	35½	17.172	49¾	40¼	17.191	45¾	44¼	17.190	42¼	47¾	17.175	39¼	50¾
17.200-40	17.206	76¾	13¼	17.214	66½	23½	17.220	59¾	30¼				17.236	50	40									
17.240-80	17.258	77½	12½	17.278	67	23	17.263	60	30	17.244	54¾	35¼				17.264	46	44	17.272	42½	47½	17.269	39½	50½
17.280-20	17.290	78	12	17.310	67¼	22¾	17.306	60¼	29¾	17.298	55	35	17.300	50¼	39¾									
17.320-60	17.321	78½	11½	17.341	67½	22½	17.349	60½	29½	17.350	55¼	34¾				17.337	46¼	43¾	17.354	42¾	47¼	17.359	39¾	50¼
17.360-00	17.380	79½	10½	17.373	67¾	22¼	17.392	60¾	29¼				17.361	50½	39½									

Approximate Lead	Gear on Worm 86, 1st on Stud 28, 2nd on Stud 44, Gear on Screw 72	Angle to Set Spiral Head (Degrees)	Angle to Set Vertical Att. (Degrees)	Gear on Worm 100, 1st on Stud 28, 2nd on Stud 48, Gear on Screw 86	Angle to Set Spiral Head (Degrees)	Angle to Set Vertical Att. (Degrees)	Gear on Worm 86, 1st on Stud 32, 2nd on Stud 44, Gear on Screw 56	Angle to Set Spiral Head (Degrees)	Angle to Set Vertical Att. (Degrees)	Gear on Worm 72, 1st on Stud 28, 2nd on Stud 56, Gear on Screw 64	Angle to Set Spiral Head (Degrees)	Angle to Set Vertical Att. (Degrees)	Gear on Worm 64, 1st on Stud 40, 2nd on Stud 72, Gear on Screw 48	Angle to Set Spiral Head (Degrees)	Angle to Set Vertical Att. (Degrees)	Gear on Worm 100, 1st on Stud 64, 2nd on Stud 72, Gear on Screw 44	Angle to Set Spiral Head (Degrees)	Angle to Set Vertical Att. (Degrees)	Gear on Worm 100, 1st on Stud 72, 2nd on Stud 86, Gear on Screw 44	Angle to Set Spiral Head (Degrees)	Angle to Set Vertical Att. (Degrees)	Gear on Worm 86, 1st on Stud 40, 2nd on Stud 64, Gear on Screw 48	Angle to Set Spiral Head (Degrees)	Angle to Set Vertical Att. (Degrees)
17.400-40	17.403	68	22	17.434	61	29	17.403	55½	34½	17.423	50¾	39¼	17.410	46½	43½	17.437	43	47						
17.440-80	17.464	68½	21½	17.476	61¼	28¾	17.454	55¾	34¼										17.449	40	50	17.450	37½	52½
17.480-20	17.494	68¾	21¼	17.518	61½	28½	17.506	56	34	17.486	51	39	17.480	46¾	43¼	17.518	43¼	46¾						
17.520-60	17.524	69	21	17.559	61¾	28¼	17.558	56¼	33¾	17.547	51¼	38¾	17.552	47	43				17.539	40¼	49¾	17.550	37¾	52¼
17.560-00	17.581	69½	20½																					
17.600-40	17.639	70	20	17.600	62	28	17.609	56½	33½	17.609	51½	38½	17.624	47¼	42¾	17.600	43½	46½	17.630	40½	49½			
17.640-80	17.666	70¼	19¾	17.641	62¼	27¾	17.660	56¾	33¼	17.670	51¾	38¼										17.649	38	52
17.680-20	17.694	70½	19½	17.681	62½	27½	17.710	57	33				17.695	47½	42½	17.680	43¾	46¼						
17.720-60	17.746	71	19	17.721	62¾	27¼				17.730	52	38							17.720	40¾	49¼	17.746	38¼	51¾
17.760-00	17.774	71¼	18¾	17.761	63	27	17.760	57¼	32¾	17.791	52¼	37¾	17.766	47¾	47¼	17.760	44	46						
17.800-40	17.800	71½	18½	17.839	63½	26½	17.809	57½	32½				17.836	48	42				17.810	41	49			
17.840-80	17.851	72	18	17.878	63¾	26¼	17.858	57¾	32¼	17.850	52½	37½				17.840	44¼	45¾				17.845	38½	51½
17.880-20	17.900	72½	17½	17.915	64	26	17.908	58	32	17.910	52¾	37¼	17.907	48¼	41¾				17.899	41¼	48¾			
17.920-60	17.950	73	17	17.953	64¼	25¾	17.955	58¼	31¾							17.920	44½	45½				17.942	38¾	51¼
17.960-00	17.997	73½	16½	17.991	64½	25½				17.970	53	37	17.975	48½	41½				17.989	41½	48½			
18.000-40	18.020	73¾	16¼	18.029	64¾	25¼	18.004	58½	31½	18.029	53¼	36¾				18.000	44¾	45¼						
18.040-80	18.043	74	16	18.065	65	25	18.051	58¾	31¼				18.045	48¾	41¼				18.077	41¾	48¼	18.040	39	51
18.080-20	18.088	74½	15½	18.102	65¼	24¾	18.100	59	31	18.088	53½	36½	18.113	49	41	18.080	45	45						
18.120-60	18.130	75	15	18.139	65½	24½	18.147	59¼	30¾	18.146	53¾	36¼				18.158	45¼	44¾				18.137	39¼	50¾
18.160-00	18.171	75½	14½	18.174	65¾	24¼	18.194	59½	30½				18.181	49¼	40¾				18.163	42	48			
18.200-40	18.211	76	14	18.210	66	24				18.203	54	36				18.236	45½	44½				18.235	39½	50½
18.240-80	18.270	76¾	13¼	18.245	66¼	23¾	18.240	59¾	30¼	18.260	54¼	35¾	18.250	49½	40½				18.252	42¼	47¾			
18.280-20	18.308	77¼	12¾	18.280	66½	23½	18.288	60	30	18.319	54½	35½	18.318	49¾	40¼	18.313	45¾	44¼						
18.320-60	18.326	77½	12½	18.349	67	23	18.332	60¼	29¾										18.340	42½	47½	18.330	39¾	50¼
18.360-00	18.360	78	12	18.382	67¼	22¾	18.379	60½	29½	18.375	54¾	35¼	18.385	50	40	18.390	46	44						
18.400-40	18.410	78¾	11¼	18.416	67½	22½	18.423	60¾	29¼	18.431	55	35							18.427	42¾	47¼	18.426	40	50
18.440-80	18.440	79¼	10¾	18.450	67¾	22¼	18.469	61	29				18.452	50¼	39¾	18.470	46¼	43¾						
18.480-20	18.486	80	10	18.482	68	22	18.513	61¼	28¾	18.488	55¼	34¾							18.513	43	47			
18.520-60				18.546	68½	21½	18.558	61½	28½	18.543	55½	34½	18.520	50½	39½	18.545	46½	43½				18.520	40¼	49¾
18.560-00				18.579	68¾	21¼				18.598	55¾	34¼	18.585	50¾	39¼									

LEADS FROM 18.600 TO 20.100

Approximate Lead	Gear on Worm 86 / 1st on Stud 32 / 2nd on Stud 44 / Gear on Screw 56	Angle to Set Spiral Head (Degrees)	Angle to Set Vertical Att. (Degrees)	Gear on Worm 72 / 1st on Stud 28 / 2nd on Stud 56 / Gear on Screw 64	Angle to Set Spiral Head (Degrees)	Angle to Set Vertical Att. (Degrees)	Gear on Worm 64 / 1st on Stud 40 / 2nd on Stud 72 / Gear on Screw 48	Angle to Set Spiral Head (Degrees)	Angle to Set Vertical Att. (Degrees)	Gear on Worm 100 / 1st on Stud 64 / 2nd on Stud 72 / Gear on Screw 44	Angle to Set Spiral Head (Degrees)	Angle to Set Vertical Att. (Degrees)	Gear on Worm 100 / 1st on Stud 72 / 2nd on Stud 86 / Gear on Screw 44	Angle to Set Spiral Head (Degrees)	Angle to Set Vertical Att. (Degrees)	Gear on Worm 86 / 1st on Stud 40 / 2nd on Stud 64 / Gear on Screw 48	Angle to Set Spiral Head (Degrees)	Angle to Set Vertical Att. (Degrees)	Gear on Worm 86 / 1st on Stud 32 / 2nd on Stud 64 / Gear on Screw 56	Angle to Set Spiral Head (Degrees)	Angle to Set Vertical Att. (Degrees)	Gear on Worm 72 / 1st on Stud 32 / 2nd on Stud 64 / Gear on Screw 44	Angle to Set Spiral Head (Degrees)	Angle to Set Vertical Att. (Degrees)
18.600-50	18.645	62	28							18.622	46¾	43¼	18.600	43¼	46¾	18.617	40½	49½						
18.650-00	18.688	62¼	27¾	18.653	56	34	18.651	51	39				18.686	43½	46½				18.697	37½	52½	18.655	34¾	55¼
18.700-50	18.730	62½	27½	18.709	56¼	33¾	18.718	51¼	38¾	18.700	47	43				18.711	40¾	49¼						
18.750-00	18.773	62¾	27¼	18.763	56½	33½	18.782	51½	38½	18.775	47¼	42¾	18.770	43¾	46¼							18.772	35	55
18.800-50	18.815	63	27	18.817	56¾	33¼	18.849	51¾	38¼							18.806	41	49	18.803	37¾	52¼			
18.850-00	18.897	63½	26½	18.870	57	33				18.850	47½	42½	18.858	44	46							18.889	35¼	54¾
18.900-50	18.938	63¾	26¼	18.924	57¼	32¾	18.913	52	38	18.925	47¾	42¼	18.942	44¼	45¾	18.900	41¼	48¾	18.909	38	52			
18.950-00	18.978	64	26	18.977	57½	32½	18.979	52¼	37¾							18.996	41½	48½						
19.000-50	19.019	64¼	25¾	19.029	57¾	32¼	19.040	52½	37½	19.000	48	42	19.028	44½	45½				19.013	38¼	51¾	19.005	35½	54½
19.050-00	19.059	64½	25½	19.081	58	32				19.076	48¼	41¾				19.089	41¾	48¼						
19.100-50	19.138	65	25	19.132	58¼	31¾	19.104	52¾	37¼				19.111	44¾	45¼				19.120	38½	51½	19.121	35¾	54¼
19.150-00	19.177	65¼	24¾	19.184	58½	31½	19.168	53	37	19.150	48½	41½	19.196	45	45	19.180	42	48						
19.200-50	19.214	65½	24½	19.235	58¾	31¼	19.230	53¼	36¾	19.222	48¾	41¼							19.223	38¾	51¼	19.237	36	54
19.250-00	19.252	65¾	24¼	19.285	59	31	19.293	53½	36½	19.296	49	41	19.279	45¼	44¾	19.272	42¼	47¾						
19.300-50	19.328	66¼	24	19.337	59¼	30¾													19.329	39	51			
19.350-00	19.365	66½	23½	19.387	59½	30½	19.356	53¾	36¼	19.370	49¼	40¾	19.362	45½	44½	19.366	42½	47½				19.352	36¼	53¾
19.400-50	19.400	66¾	23¼	19.437	59¾	30¼	19.417	54	36	19.441	49½	40½	19.444	45¾	44¼				19.431	39¼	50¾			
19.450-00	19.473	67¼	22¾	19.485	60	30	19.478	54¼	35¾							19.458	42¾	47¼				19.468	36½	53½
19.500-50	19.509	67½	22½	19.535	60¼	29¾	19.540	54½	35½	19.513	49¾	40¼	19.528	46	44				19.537	39½	50½			
19.550-00	19.578	68	22	19.582	60½	29½				19.586	50	40				19.550	43	47				19.580	36¾	53¼
19.600-50	19.647	68½	21½	19.631	60¾	29¼	19.600	54¾	35¼				19.610	46¼	43¾	19.641	43¼	46¾	19.640	39¾	50¼			
19.650-00	19.680	68¾	21¼	19.680	61	29	19.660	55	35	19.658	50¼	39¾	19.690	46½	43½							19.695	37	53
19.700-50	19.714	69	21	19.728	61¼	28¾	19.720	55¼	34¾	19.729	50½	39½				19.732	43½	46½	19.741	40	50			
19.750-00	19.779	69½	20½	19.774	61½	28½	19.780	55½	34½				19.771	46¾	43¼									
19.800-50	19.842	70	20	19.820	61¾	28¼	19.839	55¾	34¼	19.800	50¾	39¼				19.822	43¾	46¼	19.843	40¼	49¾	19.808	37¼	52¾
19.850-00	19.873	70¼	19¾	19.866	62	28	19.898	56	34	19.870	51	39	19.853	47	43									
19.900-50	19.905	70½	19½	19.912	62¼	27¾				19.940	51¼	38¾	19.934	47¼	42¾	19.914	44	46	19.945	40½	49½	19.922	37½	52½
19.950-00	19.964	71	19	19.959	62½	27½	19.956	56¼	33¾															
20.000-50	20.025	71½	18½	20.003	62¾	27¼	20.013	56½	33½	20.010	51½	38½	20.014	47½	42½	20.003	44¼	45¾	20.049	40¾	49¼	20.035	37¾	52¼
20.050-00	20.054	71¾	18¼	20.092	63¼	26¾	20.071	56¾	33¼	20.070	51¾	38¼	20.094	47¾	42¼	20.092	44½	45½						

www.ingramcontent.com/pod-product-compliance
Lightning Source LLC
Chambersburg PA
CBHW021433180326
41458CB00001B/249

* 9 7 8 1 4 2 7 6 1 4 0 7 0 *